Science
for a Green New Deal

Science
for a
Green
New Deal

CONNECTING
CLIMATE,
ECONOMICS,
AND SOCIAL JUSTICE

ERIC A. DAVIDSON

Johns Hopkins University Press

BALTIMORE

Johns Hopkins University Press
2715 North Charles Street
Baltimore, Maryland 21218-4363
www.press.jhu.edu

Library of Congress Cataloging-in-Publication Data

Names: Davidson, Eric A., author.
Title: Science for a green new deal : connecting climate, economics,
and social justice / Eric A. Davidson.
Description: Baltimore : Johns Hopkins University Press, 2022. |
Includes bibliographical references and index.
Identifiers: LCCN 2021049080 | ISBN 9781421444345 (hardcover) |
ISBN 9781421444352 (ebook)
Subjects: LCSH: Environmental quality. | Green New Deal. |
Sustainability. | Social justice.
Classification: LCC GE170 .D368 2022 | DDC 333.72—dc23/eng/20211214
LC record available at https://lccn.loc.gov/2021049080

A catalog record for this book is available from the British Library.

Special discounts are available for bulk purchases of this book. For more information,
please contact Special Sales at specialsales@jh.edu.

To the children and young adults of the twenty-first century,
who will have the means and the willpower to correct
my generation's errant ways

CONTENTS

Having enough food to eat has always been a crucial challenge for the human species. Indeed, it is the perpetual imperative for all animal species. Yet our efforts to procure or produce food have substantial consequences for the natural world in which we live. Species are being made extinct. Diverse forest and prairies have been converted to fields of monoculture. Runoff from croplands and livestock degrades streams, lakes, and even parts of the coastal ocean. Food production remains a global conundrum for society, with the world's population already nearly 8 billion and likely to hit at least 10 billion within 40 years.

Eric Davidson and I share the experience of having spent a significant part of our careers studying the impacts of agriculture on Earth's ecosystems and the services that these ecosystems provide for humans. He has conducted his research on forests and fields, and I conducted mine in coastal waters. We share a common background, having been born and raised in the great Mississippi River basin close to where our European ancestors had settled generations earlier. He spent his youth along the river system's most distant headwaters in Montana, and I grew up near its very mouth in New Orleans.

We were brought together only well into our careers when, in 2015, I recruited Eric to become director of the Appalachian Laboratory at the University of Maryland Center for Environmental Science, which I headed. I developed great respect for

Eric as a careful, thoughtful, and rigorous scientist and mentor who is exceptionally well respected by his peers—a scientist's scientist.

In *Science for a Green New Deal*, Eric Davidson goes far beyond the "sticking to one's knitting" that characterizes the traditional norm of the culture of disciplinary science. Instead, he delves into economics, human health, ethics, social justice, and even politics. I am not surprised by the book's depth, but I am impressed by the breadth of its scholarship and the clarity of its exposition. He offers a hopeful perspective on how we can use science to address concerns about our environment while also ensuring health, prosperity, and justice for all humans.

Davidson offers his call to "get to work" in full recognition that we have entered the age of the Anthropocene, with its pervasive human influence on cycles of energy, water, carbon, and other elements. Enlarging on the Pottery Barn Rule, we broke it, so we must fix it. We can do this, he argues, if we are smart and deliberate and don't just continue to muddle through.

For no issue is concerted action more important than the imperative to limit climate change. This is not only the grand challenge of the century but perhaps of human existence as a species. Rapid, human-caused climate change disrupts and makes more difficult our efforts to feed people and preserve natural diversity and productivity. It increasingly puts large numbers of people in harm's way. Davidson presents for the lay reader how scientists, trained to be a very skeptical lot, accumulated understanding of how emissions of carbon dioxide and other greenhouse gases would warm our planet Earth. By the late 1970s a rough consensus had already emerged—even Exxon knew it—although many more years of research were needed before the nations of the world would adopt the United Nations Framework Convention on Climate Change in 1992. As greenhouse gas emissions continue to steadily grow, a much more quantitative and confident appraisal has garnered virtually unanimous support among the experts.

Deep decarbonization requires phasing out burning of most fossil fuels over the next few decades. Commitment to this objective is manifest in the numerous national pledges of "net-zero by 2050" in the run-up to the recent United Nations climate conference in Glasgow. While this rapid but necessary transition poses hurdles for feeding a growing world population eager to improve nutrition and well-being, it also presents opportunities to provide food in a way that captures carbon and regenerates soils. If we think carefully, this critical transition can also provide opportunities for employment, rural development, and improved quality of life for billions of people.

Davidson was already working on this book in 2019, intending to build it around the notion of the convergence of natural and social science in the search for solutions to the complex and daunting challenges we face at this moment in our history. Such convergence provides the means to resolve environmental threats in a way that improves, not exacerbates, social justice. These are not separate issues but must be solved together.

When confronted by the annus horribilis of the coronavirus pandemic, stark incidents of racial injustice, and the extreme political polarization that ensued, Davidson, like so many of us, was forced to pause and to reconfigure his constructs. Undoubtedly, this caused him to consider more deeply society's use of science in confronting crises as well as the need for greater equity, diversity, and inclusion. He writes about these not in a moralizing manner, but with a narrative empathetic with the nontechnical reader.

Davidson chose the Green New Deal as the homily for his central premise of the need for unified environmental, economic, and social solutions. Championed by progressive politicians such as Senator Bernie Sanders and Representative Alexandria

Ocasio-Cortez, it is an array of goals and approaches to public policies that would address climate change, along with achieving other social objectives such as stimulating job creation and reducing economic inequality. I wondered why Davidson would focus on the elements of the Green New Deal resolution, considering what a lightning rod it has become for those who brand it as socialistic, economy-killing, and dead on arrival. However, in reading his book, it becomes clear that he considers the resolution not as a definitive blueprint but as an aspirational statement, more like a Request for Proposals to which we should creatively respond, which he does with each chapter.

This is not at all far-fetched, as the European Commission—the executive branch of the European Union—has formulated the European Green Deal, which enjoys wide public and political support. It includes carbon tariffs, emissions trading, the circular economy, farm-to-fork food policies, reducing fossil fuel subsidies, strategies for sustainable mobility, protections for biodiversity and forests, and forward-leaning research and innovation programs. The European Green Deal provides public support and incentivizes private investments to eliminate pollution, make the built environment more efficient, and develop clean hydrogen. It includes the collection and analysis of extensive statistics to ensure accountability and relentlessly improve effectiveness. Virtually all of these topics are, by the way, addressed in Davidson's *Science for a Green New Deal*.

Eric Davidson covers all of this with a clarity that makes his book readily approachable by the non-scientist. He tells engaging stories, often using his own life experiences from his childhood in Montana, to his days as a Peace Corps volunteer in Africa, to conducting research on deforestation in Brazil, to undergoing open-heart surgery. He provides sources of inspiration and hope. In doing so, he is one of a handful of leading scientists who is writing to offer an appropriately alarmed but optimistic outlook on the climate crisis. However, his scope is wider-reaching; he addresses the perils and promise of the human condition as a whole, making this a worthwhile read for all.

Donald F. Boesch is a professor of marine science at the University of Maryland Center for Environmental Science; he was the center's president from 1990 to 2017. From 2006 to 2017 he concurrently served as Vice Chancellor for Environmental Sustainability for the University System of Maryland.

The year 2020 was like no other in my lifetime and probably no other in yours. As we sheltered at home early in the coronavirus pandemic, we learned new social and working skills in our attempts to carry on with life under lockdown. We caught fleeting glimpses on television of how nature could bounce back once the human economy got out of the way—wildlife, from foxes to elephants, dared to venture onto roadways temporarily abandoned by humans. People accustomed to chronic air pollution in their cities began to see distant mountain ranges, often for the first time in their lives, as the air cleared of its relentless inputs of pollution from smokestacks and tailpipes.

That novelty was soon overshadowed as we came to realize that the pandemic would drag on for some time and that the poor were the most vulnerable to the mounting health and economic impacts. This should not have been a surprise, given that there was already abundant evidence, described in this book, that the poor are most vulnerable to the health effects of air and water pollution, economic recessions, natural disasters, and climate change.[1] These are often minority communities, or more informatively called "minoritized" communities, because their status has less to do with their numbers than with the historical withholding of the rights, opportunities, and privileges enjoyed by the majority.[2] In crowded housing, where social distancing was not an option, the pandemic starkly brought this vulnerability home. Likewise, frontline workers driving buses, restocking grocery store shelves, and mopping hospital floors had to keep working despite the risks of exposure to the virus.

Our focus on the pandemic was interrupted by the brutal killing of George Floyd in May 2020. Black men and women have been killed by vigilantes and police for centuries, and all killings are brutal. This death, however, was vividly captured on a nine-minute cell phone video that quickly ricocheted around the globe, bringing Floyd's last calls for breath and mercy undeniably before our eyes and ears, over and over. For many, this video and others replayed the horrors and fears that have always been their lived experiences. For others, it was a wake-up call that we, no matter how well intentioned, have been complicit in tolerating the systemic racism that enabled the graphically nonchalant killing of George Floyd. Black Lives Matter protests followed all over the world in an unprecedented expression joined by minoritized and majority communities for shared values of social justice.

As if a global pandemic and a profound social reckoning were not enough, 2020 was also a memorable year of devastating fires exacerbated by unusually long and severe droughts (2021 was equally memorable in this regard). Climate change is making such extreme events more common,[3] and nature carries on its response to climate change independent of any pandemic crisis. The fires wreaked havoc on ecosystems in Australia, the American West, and the Brazilian Amazon and Pantanal. The fires' devastation also wreaked havoc on the people, whose resilience had already been compromised by the pandemic. Not only were they tired from experiencing one devastating event after another; we have since learned that inhalation of smoke from fires actually increases the risk of getting sick and dying from COVID-19.[4]

By the time that the 2020 US presidential election came around in November, it seemed almost like an anticlimax. Nevertheless, pandemic-weary Americans voted in record numbers, by mail and in person. Perhaps they did so because so much was at stake or perhaps because the year's events just seemed to demand it.

But, alas, there was no rest for the weary, even after the long election season. No sooner had we bid good riddance to 2020 and the results of the election were about to be formally validated by the US Congress on January 6, 2021, when a mob stormed the US Capitol Building in a violent attempt to undo a democratic election.

I have always bought into the saying that a crisis is a terrible thing to waste, but can we really find silver linings and new opportunities from such an astounding series of profound crises that transpired within such a short time? We must.

Those multiple environmental, economic, health, and social stresses of 2020 and 2021, which continue today, demonstrated the importance and relevance of a pre-pandemic concept that links them together. Remembering back to "normal" times in 2019, a new idea was proposed that got limited traction at the time beyond its original political group of supporters. It was named the Green New Deal (GND), calling for legislative action similar in magnitude to President Franklin Roosevelt's New Deal response to the Great Depression of the 1930s. The addition of the adjective—*green*—suggested that, this time, environmental protection, especially the profound and urgent existential threat of human-caused climate change, would be a central focus of this new deal. Looking a little deeper, the GND's links to the environment, economics, and social justice are explicit. The intent was not simply to fight climate change and other environmental concerns, but rather to address how climate change affects the economy, the vulnerable, and social justice. Unlike any previous environmental legislation, the GND, proposed as US House Resolution 109 on February 7, 2019,[5] went well beyond water and air pollution, parks and forests, climate and biodiversity. Rather, it included the economic and social well-being of people living within their environments. Indeed, categorizing the GND as an environmental initiative is a misnomer; its economic and social components deserve equal billing.

The idea was not entirely new, as naturalists, philosophers, academicians, and even policy makers have been writing for decades about the dependency of human well-being on the quality of their surrounding environments. In retrospect, however, the GND initiative of 2019 was an unprecedented call to establish vision and goals for meaningful legislation that would connect the dots between environmental quality, economic well-being, and human dignity. It acknowledged the need for policy to be informed by good natural science, economics, and social science and to include participation by a broad spectrum of stakeholders, many of whom had been left out of decision-making and economic opportunity in the past. Although entirely relevant to the health, justice, and environmental events of 2020, discussion of the GND largely got lost among those tumultuous events. To the extent that it was discussed during the US election campaign, it was usually lambasted, with little true understanding of what it actually stood for, as being too radical and infeasible—an argument that continues today.

The GND is not the only idea of the pre-pandemic world that is reemerging as even more relevant today. As 2019 drew to a close, I was working on a book manuscript about an exciting transdisciplinary trend that is drawing natural scientists and social scientists together to search for solutions to the vexing problems facing society, such as climate change, poverty, health, and justice. This transdisciplinary approach was defined in 2014 by a US National Research Council report as "convergence research."[6] It pushes scientists of all types beyond the comfort zones of their disciplinary silos and into a world where multifaceted problems require people with diverse perspectives and expertise to work together.

As a natural scientist, I was keenly aware of the accumulating evidence from shrinking ice sheets, fires, floods, and extreme weather events, demonstrating that climate change is a very real and existential threat to humankind, and that it is occurring at an accelerating pace.[7] Although social science is not my area of expertise, I was aware that climate change is having the most severe impacts on the poor, who are the most vulnerable, often live in the most threatened locations, and have the fewest means to adapt. I noted in the late 2010s that increasingly authoritarian governments throughout the world, which were busy suppressing democratic institutions in Hong Kong, the Philippines, Thailand, the United States, Brazil, Venezuela, Hungary, Turkey, Poland, and Belarus, to name a few, often denied or downplayed the importance of climate change and other environmental or social problems that might undermine their efforts to manufacture legitimacy and authority. I began to wonder: How are climate change, social injustices, and erosion of democracy linked? Each of these issues, like many vexing problems that society faces, is bigger and more complex than any individual from a single disciplinary background can hope to understand, until we start thinking outside our disciplinary silos.

The whirlwind of 2020 caused me to pause my writing project, and to listen and reflect. With the perspective of a year or more, we can now see that the events of 2020 and 2021 have catapulted us to a realization that the comfort zones of the past will not be part of our futures. A profound shift is underway as the world responds simultaneously to a potent and persistent viral pandemic, to the visually explicit horror of racial injustice unmasked from its long-suppressed societal consciousness, and to the relentless and accelerating pace of climate change. I sense the time is now ripe to seize the moment and to embark on a new, more positive trajectory based on a new way of doing things. Convergent knowledge from many disciplines and diverse stakeholders can lead us to those new pathways, where the false choice between jobs and the environment is overturned, and where the intrinsic linkages between environment, health, and justice receive overdue recognition.

Science will be needed for this convergence, yet scientists and scientific institutions are not immune from these same engrained racial injustices. Indeed, western academia was built upon colonial hierarchical models that discouraged inclusion of newcomers who did not look or think like those who came before. So science, including the new trend of convergence research, must navigate two fronts: first, looking inward to change personal habits and attitudes and institutional cultures that no longer serve, and, second, focusing outward to engage with diverse stakeholders in community-based, context-rich, solution-seeking partnerships.

The vulnerabilities and disparities now laid bare by the events of 2020 are not limited to who has access *today* to resources and opportunities. Future generations will also grapple with fewer viable options for their well-being under a climate generally less favorable for agriculture and human health. The COVID-19 pandemic is unlikely to be a one-off event, and our resilience to future pandemics will depend upon natural and economic resources and their efficient distribution. These warnings are clear, but unless we take deliberate steps, the current disparity of resources that leaves so many people vulnerable to pandemics and climate change is likely to perpetuate itself. Increasing disparities will further separate us into those whose lives are expendable, whether by a chokehold or by a choking climate, and those who can shelter with the aid of inherited systemic privileges.

Reenter the Green New Deal. In addition to its environmentally oriented "greenness," it acknowledges the linkages between wealth and health and between wellness and justice. It envisions how those linkages can collectively define a future that is more equitable, enriching, and sustainable. It reimagines a socio-enviro-economic system that is based on shared values and informed by a convergence of knowledge and discovery from the natural sciences, social sciences, and economics, and from co-learning and co-production of knowledge with the communities with whom scientists engage. It uses the power of technology, with appropriate safeguards and guardrails, to enable sustainable and broadly accessible economic prosperity. We do not know yet how to achieve all of these lofty goals simultaneously, and we will no doubt make many mistakes and experience disappointments along the way. Nevertheless, we know a lot about how things work—in nature, in national and global economies, and in local communities—and so it is time to "converge" and to put that knowledge to work.

The GND and similar initiatives in Europe provide a starting point for simultaneously searching for policy solutions to environmental, economic, and justice issues. It is likely to continue to evolve as our understanding of the issues and potential solutions improves and as political climates shift. This book is not about utopias or lifestyles imposed by central governments on their citizens. Frankly, governments do not know the answers to our most vexing problems. Rather, it is about guiding examples of how converging natural sciences, economics, social sciences, and lessons from practical experience and knowledge of diverse groups of stakeholders can co-produce the knowledge needed to find answers leading to a more sensible and prosperous future. Going beyond the politics of a specifically named GND resolution in the US Congress, this book is about the needed green new deal thinking that will inform future governmental policies, private sector opportunities and responsibilities, structures of institutions and civil society, and individual choices.

Chapter 1 proposes a need to be deliberate about, as opposed to muddling through, the challenges we face in the current era. We now call this era the Anthropocene, due to the immense and pervasive influences of humans on our planet and on its cycles of energy, water, carbon, and other elements. Green new deal thinking can help illuminate deliberate paths through the Anthropocene.

Chapter 2 explores the links between ecology and economics and among wealth, health, wellness, and justice. It chronicles the historical patterns and injurious effects of wealth disparity on the environment, the economy, and people. Those disparities need not be inevitable.

Chapter 3 establishes the fundamental driving force of human population growth on all stressors of the environment, the economy, and social justice, using water as a guiding example.

In addition to provisioning with water resources, the growing human population must be fed. Chapter 4 lays out green new deal thinking for how regenerative agriculture, but not necessarily entirely organic agriculture, could meet that most fundamental of human provisioning challenges.

Growing food, provisioning water, and just about everything else will be made more challenging by climate change, an existential threat exceeded in consequence only by thermonuclear war. Chapter 5 describes the feasibility of how we can and must phase out fossil fuels, also known as deep decarbonization, to avoid the worst of ongoing climate change while benefiting the economy and social justice.

The profound decarbonization needed will necessitate technological transitions, many of which are already being driven by automation, and which will leave some workers worse off. Chapter 6 relates those ongoing transitions, such as the current decline of the coal industry, to the early industrial revolution in the eighteenth century and the transition faced by the Luddites. It reemphasizes the need to plan for embracing the good while mitigating the bad of technological transitions.

The most exciting and challenging of those transitions is toward a circular economy, outlined in chapter 7, where product reuse is designed at the outset of its production. The current nearly ubiquitous use of poorly recycled plastic is offered as an example of the current linear economy, in which products are used once, a small fraction is recycled, and all is eventually thrown away. The linear economy has led us to unprecedented pollution of various kinds, including plastics and climate-changing greenhouse gases, each of wicked proportions. Technological innovation will drive the circular economy of the future, with essential roles for both public and private sectors. Hence, this chapter also explores how circular economies will also require the engines of "green capitalism," which, by the way, is not an oxymoron.

All of these aspects of green new deal thinking will require a convergence of the natural sciences, social sciences, and economics. It will also require opening up academia and research institutions to foster and reward a more diverse assemblage of scientists and practitioners who seek answers to problems facing their communities. This will be a challenge for the academy, requiring a major cultural shift. Chapter 8 describes the changes of attitudes and structures that will be required in academic and research institutions to embrace justice, equity, diversity, and inclusion (JEDI). Meaningful progress on JEDI issues will help enable the needed transdisciplinary and solutions-based science to support green new deal thinking and community engagement.

Finally, in chapter 9, I share my sources of inspiration and hope.

Near the end of each of chapters 2 through 8, I quote the goals and proposed initiatives of the aspirational House Resolution 109 that are relevant to each chapter's topics. You will see that H. Res. 109 is long on aspirations and short on specifics. Its authors acknowledged this, referring to it as a "call for proposals." Responding to their call, each of these chapters ends with offerings of my own proposals. These recommendations are still aspirational and broad, but they are a bit more focused on initiatives that would support green new deal approaches to the big challenges described in each chapter.

Opinions differ widely about specific ideas and policies, many of which this book examines. The challenges that we face will likely require multiple and diverse approaches to solutions, from direct government regulations to market-based incentives. Finding solutions will require participation by individuals and communities searching not only for what is technically feasible but also for what is socially acceptable and widely desirable, thus converging on common ground for hope and for action. Still largely an aspiration, perhaps, but a new pathway can emerge from the profound events of 2020/2021, leading us toward a sustainable, prosperous, and just world.

Science
for a Green New Deal

1

- - - - - - - - - -

Muddling or Dealing?

"We made it through the Bronze Age, the Middle Ages, the Renaissance, and the industrial revolution without having a bunch of pointy-headed intellectuals telling us how to do it," my friend insisted. "And mankind will make it through the next era, whatever it'll be called, without plans dictated by governments," he said with satisfaction, while taking an equally satisfying swig of beer. Bruce was about twice my age, which was unusual for a Peace Corps volunteer. I fitted the more common stereotype of the naïve and idealistic twenty-something American, searching for adventure, experience, and meaning while sincerely trying to do something useful to help meet some of the needs expressed by my African village hosts. They made it clear that they wanted improved health care, and while I was not a nurse or physician, I could partner with them to create vaccinations campaigns, cleaner sources of water, and well-baby clinics. My wife worked with local midwives and trained volunteers to start a prenatal screening clinic. We were newlyweds, but our Peace Corps stint was not the usual honeymoon. At the moment, however, we were taking a break with our friend Bruce, enjoying a cold Simba beer on a hot African night while debating the world's problems.

Bruce was full of contradictions. He had devoted his professional life to mother-child health projects in developing countries financed by charities and government programs, but he clearly was no bleeding-heart liberal. Indeed, I was discovering his libertarian, laissez-faire side as we lubricated our conversation. Our worldview debate seemed especially odd in the context of the African nation of Zaire (now the Democratic Republic of the Congo, or DRC), where most people worked so hard for so little. The villagers who built a mud-brick and thatched-roof house for us to live in were, of course, less interested in our political worldview than in our ability to improve their families' health. As we made our modest contributions, the Zairian communities with whom we were partnering did have better health care, prenatal care, vaccines for their kids, and cleaner water, for a little while at least. We listened to what our village councils wanted and helped coordinate efforts to advance those goals. We attempted to transfer the know-how and infrastructure necessary to carry on these efforts after our departure, such as constructing and repairing concrete boxes around natural springs so that drinking water could be safely collected free of contamination. Since then, events sadly beyond the villagers' control, such as civil wars, political corruption, the Ebola epidemic, and now the COVID-19 pandemic, have left many citizens of the DRC and many other countries still deep in abject poverty.

Muddling through the Eras

The "eras" named by historians and "epochs" named by geologists have come and gone, but we have yet to create a world in which all or nearly all humans live with sufficient food, shelter, clothing, and dignity. Advances in technology have provided many of us a prosperous modern lifestyle (including cold beer), and, on average, people are better nourished and have longer expected lifetimes now than at any time in human history.[1] Nevertheless, about 700 million people are still malnourished, and in 2019 about 2 billion people did not have regular access to safe, nutritious, and sufficient food.[2] Progress in addressing poverty had reduced the number of people living in extreme poverty (defined as earning less than $1.90 per day) to less than 700 million in 2017, but the COVID-19 pandemic appears to have pushed the number in extreme poverty and malnourished back up by at least 100 million so far.[3] That means about one out of every ten people is hungry and lives in severe poverty. Despite improvements for the other nine out of ten, it is hard to see such middling progress as more than muddling through.

As I was entering adulthood, I had hope that the world was also entering into a new era where the muddling through of the past would no longer be good enough. Only a few million people lived during the Bronze Age, a few hundred million in the Dark Ages, about one billion at the beginning of the industrial revolution, and there were already 3 billion people on Earth the night that Bruce and I pontificated over a beer in 1980. Over 3 billion more people were added in just the forty years since, nearly doubling the global population in half of a modern human's lifetime. Most demographers project that the human population may eventually plateau at 10–12 billion by the end of this century.[4] How can that many people continue to extract what they need for their prosperity and well-being from the Earth's finite resources? Can the Earth absorb the many forms of waste, including greenhouse gases and plastic garbage, generated by so many people? Who will benefit from new technologies and who will be left behind, or worse yet, left to deal with the garbage, waste, pollution, and climate change caused predominantly by those who advance?

There was good reason for the idealism and optimism of my youth, and I would say that there is even more reason now, as we have better science, more knowledge, and more powerful technologies at our disposal. *But* despite those advances in science and technology, we still seem to be muddling through the decades, making incremental improvements in some areas and slipping in others. Therefore, perhaps it is time to consider that we need something like, well, . . . a plan . . . or a new deal?

The Anthropocene: When the Human Domain Clearly Enveloped the Earth

That new era that Bruce referred to as "whatever it will be called" has since been dubbed the "Anthropocene," and it looks like both historians and geologists might adopt the name. This means the era or epoch when the Earth has unequivocally become dominated by human activities, leaving clear and lasting evidence in the Earth's geologic history. That term was coined by the late Paul Crutzen,[5] who was one of the three scientists awarded a Nobel Prize for discovering in the 1970s that the Earth's protective layer of ozone in the stratosphere was in jeopardy due to proliferation of industrial chemicals that break down ozone, such as chlorofluorocarbons and supersonic jet exhaust. The notion that humans could produce so much of this stuff that it measurably destroys the layer of protective stratospheric ozone enveloping the entire Earth was a wakeup call for scientists as well as the public. It resulted in a plan, adopted in 1987, called the Montreal Protocol, which committed the nations of the world to phase out ozone-destroying chemicals. This is a good example of a global-scale plan working, as the "ozone-hole" is mending, despite some setbacks along the way, thanks to this international agreement.[6]

Evidence of human impacts on the Earth goes back thousands of years, such as abundant evidence that prehistoric human hunters and gatherers wiped out many species of plants and animals, much as modern humans continue to cause extinctions of plant and animal species today.[7] The invention of agriculture about 10,000 years ago and its subsequent global expansion also changed the Earth's atmospheric composition of carbon dioxide and methane, remarkably chronicled by analyses of bubbles of air trapped for many centuries in ice cores.[8]

Most scientists are a conservative, skeptical bunch (an idea that I will return to in chapter 5), and geologists, in particular, were not ready to declare a new epoch based "only" on a worldwide atmospheric impact that might be corrected and therefore be temporary. Rather, geologists insist on something permanent in the geologic record to demarcate when humans became a significant force in geological processes on Earth. There are many candidate indicators, including manufactured materials such as aluminum, plastics, and concrete that can be found in sediments, and more complex indicators like isotopes of carbon and nitrogen that leave a fingerprint-like mark of human influence.[9] The layer in the geological strata (layers of rocks, soils, and sediments) exhibiting radioactive fallout from nuclear weapons used in World War II and from subsequent weapons testing during the Cold War era is now proposed as marking the defining timestamp for human dominion over the Earth.[10] The radioactive fallout spread from pole to pole and from mountain tops to deep ocean sediments, leaving a lasting, unambiguous, telltale demarcating line of anthropogenic (human-caused) alteration of the geologic record.

The existential threat posed by nuclear weapons is still very much with us. Meanwhile back at the ranch, so to speak, we are changing the rest of the world at an unprecedented pace and scale—changing the climate, profoundly altering the movement of water, energy, carbon, nitrogen, phosphorus, and other essential elements for life, and permanently eliminating species of plants and animals.[11] The global mass of our buildings, roads, and other infrastructure now outweighs the mass of all trees and shrubs of the world's forests and shrublands.[12] The mass of plastic wastes accumulating in the oceans is projected to exceed the mass of all of the fish in the ocean by 2050.[13]

Humans have made similar changes on a smaller scale through-out history. The once prosperous and sophisticated Mayan na-tion of what is now part of Mexico and Central America, for ex-ample, exhausted its local resources, which contributed to its eventual demise.[14] The ancient Greeks and Romans completely deforested the Mediterranean region and allowed their soils to erode away, which decreased their food security, hurt their econ-omy, and weakened their resistance to invaders.[15] What these civilizations did on a local and regional scale in ancient human history, we are now doing globally. The sobering reality is that modern human activity rivals geological and planetary forces in influencing the trajectory of the Earth's future climate. Even more sobering, this means that the climate trajectory is largely in our hands,[16] so we'd better not screw it up. Will we muddle through or deal with a plan?

Harnessing Science

I was not so naïve as to think that I could solve these problems as a Peace Corps volunteer, nor was I so arrogant as to think that I knew the solutions, even after a couple of beers. I was simply searching for a way to make a contribution that would make a difference.

After returning home, I pursued a career in science, studying how soils and the way we manage them provide so many services to humankind, including fertile soils for agriculture and forestry, purifying water fit for drinking, and storing carbon as part of the Earth's climate-regulating system. I have published papers in sci-entific journals, mentored students as they pursue knowledge and experience, and written articles and books, like this one, for broader audiences. I have had the privilege to serve in a leader-ship position for a major scientific professional society. Still, I continue to search, along with many, many others, for how to help move the needle toward sustainable economic, environmen-tal, and social justice.

On the bright side, despite what seems like a worsening and increasingly depressing environmental situation, I have grown more convinced that we have the tools and ability to find solutions, if only we can muster the social and political will. The 2019 climate strikes of school children around the world demonstrated that they have the political will. Unfortunately, we don't have time to wait for them to take over the engines of government. It is up to us "responsible" grownups to heed their calls now.

Solutions will not come willy-nilly, as my friend Bruce viewed the progression of humankind through the ages. Rather, the challenges have become so large and urgent that the quest for solutions must be deliberate, using knowledge, experience, judgment, and values. The best science will be needed as part of the solution, and, fortunately, we have great science to draw upon— hence my optimism. The basic science that propelled humans to the moon in the 1960s also enabled the miniaturization of electronics and the exponential growth of computing power, which has since led to tremendous advances in engineering, navigation, and communication. Basic research during the last 70 years on the structure and functioning of DNA and RNA in the genetic codes of humans and viruses made possible the rapid development of an effective and safe vaccine against COVID-19. Our greatly improved scientific understanding of the Earth's climate system, from millions of years ago, through the ice ages, and to the present, now allows us to show unequivocally that the rapid temperature increases and changes in precipitation experienced during the last 50 years are mainly the result of our use of gas, oil, and coal.[17] We also have knowledge from engineering, economics, and social science to find workable and affordable solutions to climate change.

While these achievements in science are impressive and undeniable, I worry that Bruce's libertarianism of the 1980s, based on distrust of government, has grown into a more insidious, anti-intellectual, general distrust of science, despite its many advances. I can understand the distrust of science alone, which has also brought us the nuclear weapons and the ozone-eating chemicals that threaten us. Science, like anything else, must be tempered by human values and caution. At the same time, the tragically inadequate response to the COVID-19 pandemic in much of the United States, Brazil, India, and many other countries lays bare a chilling example of the consequences of ignoring science. In the summer of 2020, the lieutenant governor of Texas announced that he no longer needed advice from the nation's expert health professionals and that he had all the knowledge he needed to manage the pandemic, which then proceeded to accelerate exponentially in his state. Still worse, deliberate efforts to sow distrust in vaccines resulted in a huge surge of the Delta variant of the COVID-19 virus during the middle of 2021, and thus the almost entirely avoidable loss of many more tens of thousands of lives.

For many in the United States, not wearing a mask or getting vaccinated has become a symbol of one's politics, despite strong scientific evidence that these interventions significantly reduce the spread of the virus to surrounding family, friends, and colleagues. Curiously, similar pushback to mask-wearing requirements by Americans happened during the 1918 flu pandemic. How sad that a century of advances in medical science has not quelled the baseless, knee-jerk reaction of large numbers of propagandized people who mistakenly view sensible public safety protocols as an unacceptable threat to their personal liberties. In contrast, some have tragically overdosed on a medicine designed to deworm horses, following disinformation that they found and apparently believed on social media or TV. Ironically, perhaps some health care leaders during the current pandemic relied too much on medical science alone, ignoring the potential value of history and the social sciences to understand and to effectively address some of the reluctance to wear masks or to get vaccinated.

The governments and their citizens that followed medical science managed to slow the virus's spread (for example, New Zealand, France, South Korea, Taiwan, Vietnam, Denmark, Senegal), enabling a reasonable reopening of their economies and limiting the suffering of the most vulnerable. In contrast, those governments and communities that ignored medical science and prudent safety precautions reopened their economies too much and too soon, refused mask-wearing mandates, and failed to promote vaccinations, and thus paid a very dear price in both human lives and economic losses. We cannot allow a dismissal of science to succeed, as engineering and science, including the social sciences, have to be part of the search for solutions to ensure the health and sustainable resources for 10–12 billion people. The value of well-developed science to plan and carry out responses to threats as large as a global pandemic and global climate change must be heard above the cacophony of groundless misinformation and purposeful disinformation.

The Invisible Hand and Sleight of Hand to Thwart Planning

Despite a common foible that many of us share—that we tend not to plan ahead as much as we should—planning generally has a good reputation. Architects' plans are required to have adequate fire escapes, because experience shows that planning ahead for getting people out of a burning building helps save lives. Those fortunate enough to be able to put money aside for their children's college education or for their own retirement stand to benefit from the advice of financial planners to help them maximize returns on investments at acceptable levels of risk. Most plans often need to be reviewed periodically and revised. As visionary as President Franklin D. Roosevelt was in setting up the Social Security Administration in the 1930s to alleviate poverty among the elderly and disabled as part of his New Deal, he never could have predicted that a generation called baby boomers would follow another world war, and that those baby boomers would retire in large numbers in the early twenty-first century. Congress had to plan several decades in advance to keep the social security system solvent—a challenge that continues. Other plans are needed in a hurry. For example, when President John F. Kennedy announced the bold challenge to put man on the Moon by the end of the 1960s, NASA scientists and engineers had to plan intensively to design a series of Gemini and Apollo missions that would incrementally, but rapidly, develop the necessary technology and scientific knowledge for a successful moon shot. With so many good examples of its benefits, why would Bruce and others be suspicious of planning to meet the great challenges of our day brought on by rapidly increasing human population, pandemics, and climate change?

Distrust of government was a large part of Bruce's dismissal of the need to plan for government actions, and that sentiment remains widespread today. There certainly are plenty of examples of government plans going awry, just as there are many examples of good outcomes. Another important factor, however, is the late twentieth century espousal of a version of Adam Smith's eighteenth-century classical economic theory that the marketplace is the best place to provide for the common good. Unfettered competition, this version of laissez-faire economics holds, finds optimal solutions to economic challenges without anyone needing to plan for it. Many books have been written on this topic, and I wish to point out only two contrasting passages from Smith's own writing. On the one hand, Smith explained how pursuit of one's self-interest can benefit others:

> It is not from the benevolence of the butcher, the brewer, or the baker, that we expect our dinner, but from their regard to their own interest. We address ourselves, not to their humanity but to their self-love, and never talk to them of our own necessities but of their advantages.[18]

This is compelling stuff: by attending to their self-interests in making a living, these merchants provide us with food and drink, and if they compete with other butchers, brewers, and bakers, their goods will be priced fairly. On the other hand, Smith also warned that pursuing one's self-interest may work against the public interest:

> The interest of the dealers, however, in any particular branch of trade or manufactures, is always in some respects different from, and even opposite to, that of the public. To widen the market and to narrow the competition, is always the interest of the dealers. To widen the market may frequently be agreeable enough to the interest of the public; but to narrow the competition must always be against it, and can serve only to enable the dealers, by raising their profits above what they naturally would be, to levy, for their own benefit, an absurd tax upon the rest of their fellow-citizens. The proposal of any new law or regulation of commerce which comes from this order, ought always to be listened to with great precaution, and ought never be adopted till after having been long and carefully examined, not only with the most scrupulous, but with the most suspicious attention.

From his eighteenth-century vantage, Smith could not have anticipated the post-industrial rise of multinational corporations, their enormous political influence, and their near-monopolies, but his warning that political corruption and influence were likely to undermine the free market system was prescient to say the least. The "butcher, brewer, and baker" statement exemplifies Smith's view that the marketplace's "invisible hand" will find a solution that benefits all. Neoclassical economists of the twentieth century called this "general equilibrium" in their textbooks. Their theoretical and mathematical models seemed to show that competition in the marketplace would lead to optimal provisioning of goods, services, and labor and a fair distribution of wealth, assuming, of course, that competition is not corrupted by politics or other influences. Hence, they assume that Smith's warning can be mostly ignored. Economist Heather Boushey notes that these neoclassical economists came to see their role as defending the market economy by demonstrating that it is "ruled by a set of natural laws that were both self-regulating and (generally) socially beneficial."[19] In that case, who needs planning? In fact, taking neoclassical economics to its logical extension, planning might run contrary to natural law of the marketplace.

In reality, the sleight of hand of the wealthy and politically connected often thwarts the invisible hand of perfect competition, as Smith warned us it would. The earlier President Roosevelt—Theodore—both embraced free market capitalism and recognized the need for government to regulate excessive greed and to equalize the power imbalance between corporations and consumers with what he called "Square Deal" policies.[20] Yet later in the twentieth century and continuing today, a reliance on (and even devotion to) a belief in the so-called free market, or at least the current adulterated version of it, is often used to argue against government plans and actions to deal with what economists call market failures. These market failures include imperiling things important to people that are not traded effectively in the marketplace (which economists call "externalities"), such as climate, health, education, and justice.[21]

In the next chapter, we examine data demonstrating that unfounded reliance on the so-called natural laws of neoclassical economics have, instead, produced inequalities today that resemble those that Teddy Roosevelt tried to address a century earlier. The marketplace and its competition will remain a powerful force that, when assigned proper guardrails in a mixed market system, can contribute to efficiencies and prosperity, while avoiding monopolies, political favors, and perverse inequalities. The "natural laws" of the market economy, however, which some economists have invoked since the mid-twentieth century to preclude the need for meaningful government planning, are little more than a mirage. Instead, today's lessons of economics and history reveal all too clearly the need for plans to alleviate poverty, avert catastrophic climate change, and tame a global pandemic as we look for alternatives to continued unplanned muddling through the Anthropocene.

Convergence for Green New Deal Thinking

A phrase first coined by journalist Thomas Friedman in 2007 reemerged in 2019: a "Green New Deal" (GND) suggests that we need a change as profound as President Franklin Roosevelt's New Deal during the Great Depression of the 1930s.[22] Let us not forget that the Great Depression was not only a global economic disaster but also an environmental one. The Dust Bowl, caused by poorly informed farming practices that exposed bare soil, followed by severe drought and wind storms that blew the soil away, devastated the US and Canadian prairies, impoverishing already poor farmers and leading to mass migration.[23] Agronomic scientists had warned against the consequences of poor farm management but policy makers did nothing to avert the impending catastrophe. The US Soil Conservation Service (SCS) was created only after the Dust Bowl crisis was well underway.

In fact, it was not entirely a coincidence that a giant dust storm reached Washington, DC, on the very day that Congress deliberated on a bill to set up the SCS. The supporters of the bill, recognizing that a crisis is a terrible thing to waste, had learned through telegrams from the Midwest that the dust storm was blowing east, and so they deliberately postponed the hearing in Congress to the day that the storm was expected to arrive in Washington. Faced with undeniable evidence in their eyes and nostrils, Congress passed the needed legislation. The SCS was but one of a raft of initiatives that created new regulatory agencies and funding for public works and social safety nets to address the economic crisis of the Great Depression, which dramatically changed the role of the federal government.

Today the world is facing environmental, economic, and social issues as challenging as those of the Great Depression. Dust storms are common in parts of China and West Africa today. Not unlike the Okies escaping the Dust Bowl of the American Midwest in the 1930s, farmers and their families are leaving their homes in Central America in record numbers, desperately trying for a new life in the United States because environmental, social, and economic conditions have made it impossible for them to survive in their native countries. Indeed, mass migrations are occurring throughout the world, driven by combinations of economic disparities, war, drought, and environmental degradation.[24] In response to these epic challenges, major actions on a scale equivalent to the response to the Great Depression are needed, and not just in the US. Green new deal thinking and its benefits to humanity apply to communities and countries across the globe.

Enter the 2020s and we see not only the urgent need for the greenness of a new deal, but also one that is more inclusive than Franklin Roosevelt's New Deal nearly a century ago. Black Americans were mostly left out of Roosevelt's New Deal; they were locked into the Jim Crow laws of segregation and post-reconstruction era laws that prevented many of them from voting despite their Fifteenth Amendment rights—a fight that has reemerged today. The Civil Rights Act of 1964 was a major new deal with respect to formal legal rights for people of color, but three centuries of American history based on a social structure of racial inequities and injustices could not be reversed with a stroke of President Lyndon B. Johnson's pen alone. Race-based economic, social, and legal inequities have since only been nibbled away at the margins through middling, incremental progress. Thus, in 2020, the cell phone in the hands of a bystander recorded the brutal murder of George Floyd, and a cell phone in the hands of an accomplice recorded the hate-driven killing of

Ahmaud Arbery. What was new was not the disregard for Black lives; it was the cell phone as witness and recorder, creating videos, instantly ricocheted around the world, that none can deny and few can forget or shrug off.

The environmental, economic, and social justice challenges are enormous, but we know a lot about how to approach them, and we are beginning to think about those approaches in an integrative framework. Indeed, the green new deal concept proposed by US legislators in 2019 offers a framework that emphasizes how environmental problems cannot be solved in a vacuum, without being intimately linked to socio-economic, human rights, and racial justice concerns. The subsequent pandemic and the video murders of 2020 drove home further that the poor and marginalized are the most vulnerable to disease, environmental pollution, and injustice. The convergence of these challenges in the early 2020s requires a new approach to finding integrated solutions. What better time for green-new-deal thinking?

Based on a recommendation of a report of the National Research Council,[25] the National Science Foundation (NSF), sensing such a need, has embraced a new term—convergence research—which they describe as

> a means of solving vexing research problems, in particular, complex problems focusing on societal needs. It entails integrating knowledge, methods, and expertise from different disciplines and forming novel frameworks to catalyze scientific discovery and innovation.[26]

I like that description, but it should be applied beyond research. We need to search for and experiment with convergent solutions in everyday life, both now and as we envision the future.

A green new deal could be such a framework for convergence thinking and policy making. Its form is still evolving and being incorporated into various plans, such as parts of President Joseph Biden's "build back better" initiatives, the European Green Deal, and others. It will continue to evolve beyond any one administration of a single nation, taking on perspectives and experiences as it goes. It is not a hard-and-fast, detailed, comprehensive, top-down plan, imposed by elite intellectuals, like the kind that my friend Bruce ridiculed. Rather, it should be a guiding roadmap of pathways based on all participating stakeholders' values and ideas. Scientists and experts from many disciplines will play important supporting roles. Leaders from affected communities must be included, and together we will co-produce the needed new knowledge to find those hitherto elusive solutions for enabling environmental and economic sustainability and social justice. Many forks lie in the road, but the destination is clear and the course is deliberate, as we collectively explore, plan, create, and implement the new-deal future that we dare to dream for ourselves and our children.

2

No Tree, No Bee, No Honey, No Money

I regret that I did not write down the name of the African student at an international conference in Germany in 2000 whose poster bore the title that I have borrowed for this chapter. She was the daughter of African farmers in a remote village, and she told me that she had been very fortunate to be able to attend university. The students attending this conference were young agronomists from all over the world, pursuing their master's and doctoral degrees in Germany. Posters of their research projects were featured one afternoon; I was drawn to the title's cleverness right away, but its full meaning and prescience did not sink in until much later. Searching the internet, I see a few other applications of this phrase as well, but I credit the student, and I hope that wherever she is now, she and her community are prospering and that she does not mind me borrowing her insightful title for this chapter.

The people of her village scrape out a living in all sorts of ways—tilling small plots of land, selling farm produce and crafts in the local marketplace, collecting wood and water from the nearby forest, hunting and fishing, and, yes, collecting honey to eat and to sell for supplemental income. People encourage the bees by setting out structures in the trees in which the native bees can locate their hives, essentially doing a tree-based version of beekeeping (fig. 2.1). They know that if you cut down the trees to make room for more farmland, the bee population will decline, and with fewer bees, there will be less honey and one less source of money in the portfolio of means to make ends meet in their family budget.

FIGURE 2.1 Beehive Boxes Suspended in a Tree in Laikipia County, Kenya. Photo by the author

The student's poster and her master's degree research focused on how prudent management of the natural resources in her community, including integrating forest patches with croplands, could improve the resiliency and the overall economic prosperity of its inhabitants. As with any good balanced investment portfolio, multiple revenue streams, she argued, add resiliency when one of those income sources suffers, as in a crop failure in a drought year. She further investigated the resources needed for more villagers to take up profitable beekeeping, merging local knowledge from beekeepers and elders in her community with the contemporary published scientific literature. She synthesized these research findings to serve her villagers while also expanding knowledge that could be useful for other communities. Through her studies, the master's candidate was searching, like many of us, for how she could make a difference for her community and for the world.

Money Doesn't Grow on Trees (or Does It?)

When I asked for expensive things as a kid, my mother would sweetly but sternly lecture me: "Money doesn't grow on trees, honey." She was right to teach me frugality, but she was sort of wrong about the value of products that grow on or in trees.

Most Americans and Europeans think of beehives as the stacks of drawers that they see set out in farmers' fields as they drive by on the highway. The boxed beehive takes the place of the tree in the American and European farm version of this story. More importantly, the money goes well beyond the honey and extends to the value of the pollinated crops, including dozens of kinds of fruits, vegetables, and nuts that are part of our daily diets, as well as cotton for clothing and alfalfa for livestock feed. Almond plantations in California depend entirely on imported European honey bees for pollination, for which the state's almond farmers pay beekeepers some $400 million per year.[1] The return on investment usually makes the cost worthwhile, as the California almond crop is worth more than $5 billion.[2] (Mom would have been impressed by all that money growing on trees.) Unfortunately, honey bees are in decline in much of North America and Europe due to a combination of stresses not yet fully understood by scientists and beekeepers, including infection by mites, invasions of bee-killing hornets, poor nutrition, overuse of pesticides, and climate change.[3] As a consequence, farmers are forced to invest increasingly large sums of money in beekeeping to ensure that enough bees will survive to pollinate their crops.

The trees-to-money link through the bees and their honey, in both African villages and California almond plantations, is a powerful metaphor for how intertwined our ecological and economic systems of prosperity really are. Bees also perform their pollination service and honey making in the wild, where many species of bees build nests in forests, probably surviving better than in boxes at the farm edge. The economic linkages may not seem so obvious to people outside the farming community, for whom money comes out of automated teller machines, while politicians focus on job creation as the primary basis for economic prosperity. Job-creating industries are, indeed, crucial multipliers of wealth, producing goods and services and recirculating money through sales and wages.

Digging deeper, however, every job and product can be traced to some mineral, crop, forest, energy source, or water resource on which the workers and products ultimately depend. Our money and indeed nearly all forms of our wealth originate in some way from natural resources and our cleverness in extracting and husbanding those resources.[4] Destroy, foul, or deplete those resources, whether they be forests, aquifers, soils, oceans, the atmosphere, or bees, and we undermine our own future economic prosperity. Prosperous nations, like the United States, usually have some combination of fertile soil, abundant water, hospitable climate, extractable resources, and innovative work-ers. Nations without rich natural resources depend on those of their trading partners.

My mother knew all about the challenges of making ends meet in our family budget, although I doubt that she ever considered beekeeping. In any case, Mom, stuff worth a lot of money does grow on or in trees, and, more broadly, our economy and pros-perity are dependent on how cleverly we manage all of our natu-ral resources.

No Wealth, No Health, No Wellness, No Justice

The environment alone, represented metaphorically here by its trees and bees, does not create wealth, but it is an indispensable foundation. Human expressions of ingenuity, including early forms of engineering (for example, the invention of the plow and the smelting furnace), traditional knowledge (beekeeping, using fire to manage the landscape), and modern forms of science (crop breeding, biomedical science, computer science), are also essen-tial. Industrious workers and innovators need social structures that ensure a safe and healthy work environment, a fair share of the wealth that their innovations and hard work help generate, and a respectful dignity, thereby contributing to the stability and sustainability of the wealth-generating process. History has shown that there is no guarantee that workers will get their fair share or a safe work environment unless society makes a con-certed effort to make that happen.

Developing a novel approach to studying historical tax records, French economist Thomas Piketty stirred a few pots when he demonstrated in his 2014 book, *Capital in the Twenty-First Century*, that the disparities in wealth and the accompanying deplorable work conditions for the poor during the late nineteenth-century Gilded Age appear to be returning today after only a relatively brief digression in the mid-twentieth century.[5] Personal wealth, as he calculated it, came from inherited capital, such as land, business assets, and investments, and through income from employment and business profits. Inheritance was the main source of personal wealth prior to the industrial revolution, especially for European landowners, but income from the growth of new industries also helped create a new wealthy class in the late nineteenth and early twentieth centuries. From 1914 to 1945, however, the two world wars separated by the Great Depression decimated much of the wealthy classes' capital and income (fig. 2.2). The decline started with World War I in Europe and was delayed in the United States until the Great Depression.

After World War II, prosperity began to recover and wealth was more evenly distributed among US economic classes. Labor unions gained strength and were effective at bargaining for

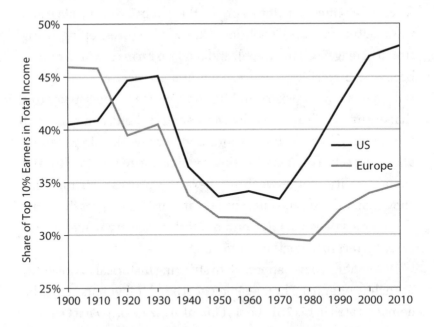

FIGURE 2.2 Income Disparities in Europe and the United States, 1900–2010. The income of the highest-paid tenth of the population as a percentage of total national incomes illustrates the history of income disparity. Redrawn by the author from publicly available data in Thomas Piketty's *Capital in the Twenty-First Century*, http://piketty.pse.ens.fr/en/capital21c2

better wages and working conditions. A progressive tax structure helped redistribute income by taxing the rich and investing public funds in education and other benefits for the middle and lower classes. Social security, enacted as part of the New Deal during the Great Depression, enabled the elderly and disabled to escape severe poverty, live out their lives in dignity, and contribute to society and to the economy.

Another good example of a policy that narrowed income disparity is the GI Bill, which enabled nearly 8 million American veterans returning home from World War II to participate in education or training programs and enabled more than 4 million vets to obtain home loans backed by the Veterans Administration. This was controversial because, prior to the war, higher education was limited primarily to the wealthy, and many members of the US Congress thought it would be too costly to extend these opportunities to the middle and lower classes. It was these investments in returning war veterans, however, that helped build the wealth of a growing middle class, my father, uncle, and our families included. Incomes rose for most white Americans and even favored the lower end of the economic spectrum. From the early 1960s until about 1980, incomes rose by 2.6% per year for the bottom half of the US population; in contrast, the top 1% percent of earners also saw their incomes rise, but by only 1% per year.[6] As a consequence of the GI Bill and other public interest policies, the income disparity gap declined for white Americans after World War II to a low in the 1970s.

Sadly, implementation of the GI Bill was uneven, and systemic barriers prevented Blacks from getting home loans and being admitted to colleges and universities, so few of the 1.2 million African American World War II veterans were able to benefit from this wealth-building and human resource–enabling policy for the American middle class.[7] Excluding most African Americans from benefiting from the GI Bill in the 1950s and 1960s extended an earlier pattern of racial discrimination by banks in the 1930s known as redlining (so called because of the practice of marking maps in red ink to highlight predominantly African American neighborhoods where institutional lending was discouraged). This systemic discrimination prevented investments in housing in African American neighborhoods and thus limited the ability of those residents to accumulate wealth as homeowners. The result is an enduring legacy with unexpected impacts, such as vulnerabilities of human health and even climate change resilience.

The same urban neighborhoods that were redlined in the 1930s and denied GI Bill loans in the 1950s are inhabited by predominantly minority residents today. They tend to be nearly 5 degrees Fahrenheit warmer on average during the summer than nearby white neighborhoods. They have more pavement and less shade, leading to more heat-related hospitalizations and emergency room visits.[8] In addition to being hotter, these neighborhoods also tend to be closer to other forms of pollution of air, water, and soils. Emerging epidemiological evidence from the COVID-19 pandemic suggests that people chronically exposed to small particles of air pollution (called PM2.5, for the 2.5-micron diameter of the particulate matter suspended as pollution in the air that penetrates deep into the lungs when inhaled), are more susceptible to falling seriously ill from the virus.[9] The higher emissions of PM2.5 from industry and traffic in low-income neighborhoods and communities of color, which tend to be located near factories and highways, may have contributed to the higher morbidity and mortality among these populations during the pandemic.[10] Whether it is in response to a knee on a throat or microscopic soot in the air, these communities can't breathe for reasons that are systemic.

Those denied economic opportunities and exposed to elevated pollution are not limited to urban communities. Among the most egregious examples of the many rural communities left out are the Native American indigenous people, historically forcibly relocated onto tribal reservations. Growing up in Billings, Montana, I lived only 80 miles from the Crow Reservation, but no Native American children attended any of my public schools. I am quite certain that the children on the reservation did not have the educational and employment opportunities that I enjoyed only a few miles away. That deliberate physical and social separation was the heritage of structural racism, based on colonialist notions that persist today. The COVID-19 pandemic has brought renewed attention to the long history of lack of adequate health care, electricity, potable water, sanitation facilities, and access to voting in tribal reservations across the country, which has left these indigenous communities more vulnerable to pandemics, economic recessions, and climate change.[11]

Some might consider the GI Bill, social security, labor rights, and other progressive New Deal and post–World War II policies as forms of creeping socialism, but these programs actually contributed mightily to the success of capitalism in our mixed economy, at least for white communities. My uncle used the GI Bill to go to college and later pursued a medical degree. My father was able to buy a home with no down payment because of the GI Bill, which enabled my parents to build up equity over time and help create the modest wealth of the middle class. My father was active in the Young Republicans and later had a career as a salesman for a private insurance company, but any suspicion that the GI Bill was part of a socialist agenda never worried him. Many blue-collar workers who belonged to labor unions also became homeowners and bolstered a growing middle class of consumers whose spending helped further stimulate the economy. The safety net of social security is embraced by its beneficiaries on both the political left and right.

A change in the global political climate during the 1980s, however, reversed policies in ways that began increasing income disparity, leading to today's astonishing wealth, income, and opportunity gaps. Tax rates for the wealthy and for corporations have been lowered several times since the 1980s, labor unions have lost much of their clout, and investments in public education have been mostly stagnant. These policies of the last 40 years reversed the middle-class wealth-building trend of 1945–1980 and increased wealth and income disparities (see fig. 2.2). This re-establishment of a Gilded Age–like income gap is more pronounced in the United States than in Europe. The richest 1% of Americans now own 39% of the country's wealth, a surprisingly rapid increase from 30% in 1990 and 24% in 1980, while the average real wage of American workers has remained flat and actually fell for the least educated workers.[12] The rising prosperity of the wealthy is due to a combination of exorbitant incomes, such as the super salaries of CEOs,[13] and a reemergence of capital as a source of wealth, as hedge fund managers and start-up success stories generate capital, and CEOs get additional compensation as stock options, which then become family inheritance.

At first blush, increasing disparity of income and wealth seems wrong mainly on the grounds of heart-wrenching suffering endured by the poor. For example, when the executive director of the World Food Programme spoke after his organization was awarded the 2020 Nobel Peace Prize for fighting hunger on a global scale during the COVID-19 pandemic, he felt compelled to plead for $5 billion more in donations to stave off the growing acute hunger provoked by the pandemic, which threatened more than 200 million people.[14] He directly appealed to both governments and the many billionaires who could come together to meet that challenge.

Hunger is not only a story about developing countries. Even before the pandemic hit, US Department of Agriculture data suggested that 35 million Americans in 13.7 million households experienced food insecurity at some point during 2019 (defined as being uncertain of having, or unable to acquire, enough food to meet the needs of all their members because they had insufficient money or other resources for food).[15] More than one-third of these households were Black or Hispanic.

The rationale for being concerned about wealth disparity goes well beyond empathy for the poor. As economist Heather Boushey describes, there are other insidious effects of growing wealth and income disparity that affect the economy for all of us.[16] First, as the rich get richer, they gain more political influence and are able to sway government policies, such as tax cuts that predominantly favor them, thus perpetuating and exacerbating the income divide. They also use their influence to reduce governmental protections for labor and the environment and to thwart campaign finance reform that might limit their future influence. Most notably, they use their political influence to facilitate corporate mergers that reduce competition—the very foundation of capitalism—just as Adam Smith warned they would (see the previous chapter). Consumers, workers, and the environment suffer.

Second, the super-wealthy spend a smaller percentage of their incomes on consumer goods and therefore do not stimulate the economy as much as would happen if that income went to people in lower income brackets, who, like my mom, spend most of what they earn to make ends meet. Some economists call this the "you can only wear so many shirts" phenomenon. Even the most expensive silk shirts and expansive closets filled with designer shoes still represent only a small fraction of the super-wealthy's income. Furthermore, their spending tends to be for high-end products, thus creating demand and promoting innovation at the high end rather than innovations for improving the less expensive products needed by the less wealthy. A good example is the all-electric 2022 Hummer Edition 1: would-be buyers, apparently with plenty of pocket change to cover the $112,595 price tag, swelled the waiting lists a year before the car actually went on the market. I confess that it looks like a really cool, off-road, luxury, all-electric pickup, designed for the new-age person, perhaps a cross between a Gilded Age robber-baron and the Marlboro man. What is really needed, however, is innovation in the electric vehicle (EV) market at the lower end of the price range to make them more affordable to a wider clientele. Unfortunately, growing income inequality means that there is less demand, and hence less motivation for innovation, for EVs at the lower end of the market.

Even after adding on yachts and private jets, there is still a lot left over in the annual budgets of the wealthy that usually gets invested in securities rather than being spent on consumer products. Excess income invested in stocks can help keep stock markets growing, demonstrating, as it did during much of 2020 and 2021, that the stock market can be insulated, at least for a while, from the vitality and volatility of the underlying economy. Once again, the rich benefit disproportionally from a growing stock market, while the economy reflects insufficient growth of good-paying jobs for the middle and lower economic classes.

In theory, if those stock market investments were providing capital for industries to expand and thereby create new jobs, this wealth effect could trickle down as new jobs are created. To some degree this happens, but Boushey and other data-driven economists of her generation show that those old theoretical assumptions mostly do not hold up to rigorous analysis. The 2017 US tax cuts, like those of previous decades, failed to pay for themselves by stimulating the economy and creating jobs in sufficiently large numbers to increase overall tax revenues, as the supply-side economists of the 1980s onward frequently and incorrectly predicted. Instead, they led to lower tax revenues, which ballooned the federal budget deficit, which also meant less support for government spending on investments in health care, education, environmental quality, and infrastructure that would increase the well-being of lower- and middle-income families and create more jobs. Hence, Boushey shows that concentration of wealth at the top actually decreases the proportion of investments in what people need, and that makes the economy more sluggish. With fewer educational and employment opportunities, lower and middle classes are also less able to contribute as clever innovators, industrious workers, small business capitalists, and generators of knowledge. A report from the Brookings Institution and the Federal Reserve Bank of San Francisco estimated that about $23 trillion of GDP was lost from 1990 to 2020 due to labor inequities related to race and ethnicity in the United States: "The opportunity to participate in the economy and to succeed based on ability and effort is at the foundation of our nation and our economy. Unfortunately, structural barriers have persistently disrupted this narrative for many Americans, leaving the talents of millions of people underutilized or on the sidelines. The result is lower prosperity, not just for those affected, but for everyone."[17]

Co-production of Knowledge

The young African woman studying in Germany exemplified the value of investments in people who become generators of knowledge. She was the beneficiary of a development project that provided her access to a master's degree, but her experience was an unusually fortunate exception to the rule. Twenty years later, most of her countrymen, and especially her countrywomen, still have few educational opportunities. In the developed world, increasing disparities of wealth and income deny the privilege of quality education to most poor people, who are predominantly (but not exclusively) people of color. Consider what a society loses when large numbers of its citizens do not have opportunities to contribute to our collective human cleverness to find solutions to challenges and to bring prosperity to our communities and nations.

The complex problems that society faces, such as how to limit or reverse climate change while maintaining a sustainably prosperous economy, obviously require cooperation among experts from a wide array of disciplines, such as engineering, economics, social science, and natural science. We have a lot of expert knowledge and innovative ideas, but we need more than just interdisciplinarity and cooperation among so-called experts. We also need creative inputs from citizens facing their everyday challenges. Finding solutions that people will embrace, from beekeeping to lightbulbs to farming practices, requires engaging communities that decide what to buy, what to use, what to value, and how to live. Those people, those everyday decision makers, coming from all creeds, ethnicities, and economic statuses, must be at the table, contributing their ideas, talents, and individual expertise to find what works.

A good example of the limitations of experts is reflected in the effort to find safer and more efficient ways of cooking in low-income countries. In many parts of the world, including the Zairian village where I served as a Peace Corps volunteer, the women still balance their cooking pots on three stones over a smoky fire. Several well-meaning experts designed efficient wood-burning cook stoves that also reduce smoke inhalation, a major cause of respiratory disease and premature death in the developing world,[18] only to be disappointed that the women doing the cooking did not adopt them widely. The late Kirk Smith, after working for decades on improved wood cook stoves to reduce indoor pollution, pivoted late in his career to recommending use of LPG gas (also known as propane), which has been readily adopted by women in India and throughout Asia and Africa when it is available.[19] Propane is a fossil fuel, so its use in cooking is not ideal from an environmental perspective, but reducing the demand for firewood is a counterbalancing environmental benefit, and eliminating harmful smoke exposure to women cooking the family meals is an overriding social concern. Propane will one day be replaced in India and other nations by rural electrification generated by renewables or by methane produced from livestock manure in biodigesters, but in any case, the social science of acceptance of technology is just as important as the technology itself.

In Europe and the United States, new types of slow-release fertilizers have been engineered in an attempt to reduce water pollution from nutrient runoff. The slow-release technology allows the crop to take up a larger fraction of the applied fertilizer over the course of the growing season, and this improved efficiency of fertilizer use should enable the farmer to apply less fertilizer overall.[20] Unfortunately, farmers have been reluctant to adopt these slow-release fertilizers, and related products, called enhanced efficiency fertilizers. Many farmers are reluctant to take a risk that applying less of this higher-cost special fertilizer will give the same crop yield and economic return as their usual practices. They are risk averse because, like the rest of us, they need to send their kids to college and want to retire in comfort. The household cooks and the farmers must be part of the teams of innovators, because their knowledge to help find and apply solutions is immensely valuable.

Similar to these examples of individuals' decision-making in the home and on the farm, solutions to global problems will also require inputs from many stakeholders. Al Gore called climate change *An Inconvenient Truth,*[21] and that title applies equally to pandemics and racism. In fact they are all connected, and "inconvenient" is an understatement for all three. We will not find solutions to climate change, health risks, or systemic racism until we address all three simultaneously. The vulnerabilities can be traced to disparities in wealth and income that have been increasing globally since the 1980s. The convergence of knowledge needed to find solutions to these vexing, intertwined challenges must come from many disciplines and from people with many relevant experiences. Hence, we need a new strategy, or a series of strategies, derived from farmers, homemakers, business leaders, enterprising capitalists, scientists, engineers, economists, community leaders, youth, and people with all sorts of personal and professional experiences, for cleverly extracting economic prosperity and good opportunities for all workers, while also maintaining the ecological integrity of the Earth.

Put another way, pigeonholing the environment, the economy, and social justice into separate solution bins is so twentieth century. Even calling them separate legs of a common stool misses the point. Rather, the challenges of environment, economy, and justice, as well as their solutions, are so intricately interwoven that only a convergence of knowledge can provide a pathway to progress. Such convergence does not happen willy-nilly; it requires some concerted effort, planning, and encouragement by the institutions of civil society, including public and private sectors, and with a significant role for government.

Simply put, how can we continue to extract from the Earth's resources what we need for the prosperity, well-being, and dignity of current and future generations of billions of people without exhausting or polluting those resources? Each expert can answer part of this question, but the question is too existentially important to leave to only economists or only ecologists or only engineers, or (for that matter) only experts. It will need all of the innovative juices and insights that diverse communities across the world can muster.

I offer evidence for my optimism in the following pages. Despite the current gloomy prognoses about climate change, economies crippled by a global pandemic, and long-festering social and racial injustices, the ingredients are coming together for a new vision of human interdependency with the Earth's support systems, including its trees and bees and the millions of other essential natural resources. This convergence will re-write the human trajectory for how the many innovators of human cleverness will co-produce knowledge-based solutions to secure environmental, economic, and social sustainability. Whether we call it green or new, or something else, it will definitely be a big deal.

Recommendations

The Green New Deal (H. Res. 109, 116th Cong., 1st Sess. [introduced February 7, 2019]) speaks to engaging multiple stakeholders to find solutions to environmental, economic, and justice issues:

> a Green New Deal must be developed through transparent and inclusive consultation, collaboration, and partnership with frontline and vulnerable communities, labor unions, worker cooperatives, civil society groups, academia, and businesses.

It also addresses inclusiveness so that everyone can contribute to the ingenuity and cleverness needed to find solutions:

> providing resources, training, and high-quality education, including higher education, to all people of the United States, with a focus on frontline and vulnerable communities, so that all people of the United States may be full and equal participants in the Green New Deal mobilization.

Contrary to misconceptions that it rejects capitalism in favor of socialism, it actually calls for a more equitable form of capitalism by curbing unfair competition and monopolies:

> ensuring a commercial environment where every businessperson is free from unfair competition and domination by domestic or international monopolies.

Although its preamble emphasizes the current state of wealth inequality,

> the greatest income inequality since the 1920s, with—(A) the top 1 percent of earners accruing 91 percent of gains in the first few years of economic recovery after the Great Recession; (B) a large racial wealth divide

amounting to a difference of 20 times more wealth between the average white family and the average black family; and (C) a gender earnings gap that results in women earning approximately 80 percent as much as men, at the median.

wealth is mentioned only once in the list of needed actions:

directing investments to spur economic development, deepen and diversify industry and business in local and regional economies, and build wealth and community ownership, while prioritizing high-quality job creation and economic, social, and environmental benefits in frontline and vulnerable communities, and deindustrialized communities, that may otherwise struggle with the transition away from greenhouse gas intensive industries.

Finally, relevant to this chapter, it endorses actions that would strengthen the lower and middle classes:

creates high-quality union jobs that pay prevailing wages, hires local workers, offers training and advancement opportunities, and guarantees wage and benefit parity for workers affected by the transition; strengthening and protecting the right of all workers to organize, unionize, and collectively bargain free of coercion, intimidation, and harassment.

Interestingly, this GND resolution is silent on the issue of taxation. Narrowing the disparities of income and wealth back to the levels of the mid-twentieth century will require new investments in education, training, and employment to lift up the lower and middle classes as described in the passages of the GND resolution quoted above. We will return to those topics in the following chapters. Regardless of whether these initiatives are paid for by government deficit spending, revenues from a stronger economy, private sector investments, higher taxes, or most likely some combination thereof, **narrowing wealth and income disparities will require that the wealthiest individuals pay a larger share of the tax burden,** as they did prior to 1980.

Specifically which tax policies would be most effective, such as increasing the top marginal income tax rate, increasing capital gains taxes, or instituting a new wealth tax, is beyond the scope of this book, but some leveling of the playing field is clearly needed. Not only should disparities be narrowed from both the bottom up and the top down for a sense of fairness, but also to mobilize resources sequestered at the top to support broader wealth-producing endeavors at the bottom and middle. As shown by Heather Bushey and other economists of her generation, not only can this can be done without inhibiting economic growth, but it can also actually stimulate the economy by transferring resources to the people who will spend them and who will build new innovative enterprises from the bottom up. As the subtitle of her book *Unbound* suggests—*How Inequality Constricts our Economy and What We Can Do About It*—she offers several additional proposals for economic policies that are consistent with a mixed economy based on capitalism and that will also narrow income and wealth disparities.[22] These include *policies for equitable growth, eliminating economic discrimination, curbing monopolies and monopsonies, and going beyond GDP to measure national income accounts that demonstrate multiple monetary and nonmonetary benefits of these policies and to quantify who gets those benefits.* Let us also include among the monetary accounts a proper valuation of the trees and the bees and wellness and justice, as they are all inextricably intertwined.

3

- - - - - - - - -

Are There Too Few
or Too Many People?

The radio was so loud on the SuperShuttle that I couldn't avoid listening to it on my way back from Boulder to the Denver airport. The day's top news item: Demographers had reported that the 300 millionth American was to be born on that day in 2006. "Is that a problem?" the talk show hosts asked. "Are we running out of space? Has the sky fallen like those environmentalists predicted it would back in the 1970s when the population explosion was their main concern?"

This question of population growth was also relevant to the scientific meeting I had been attending in Boulder, which focused on how to measure the carbon dioxide that plants take out of the air through photosynthesis, which plants and soils then release back into the air through respiration. Depending on the net balance between the carbon dioxide consumption by photosynthesis and carbon dioxide release by respiration, the world's forests could either help reduce the severity of human-caused climate change or make it worse. Lucky for us, forests and oceans have been taking up about half of the carbon emitted by humans,[1] making climate change less rapid than it would have been otherwise. As the world warms, however, forests could become net sources of carbon emission into the atmosphere because of increased rates of respiration, the limited availability of water for photosynthesis, and increased fire.[2] The number of humans burning fossil fuels and clearing forests is another big part of the climate change equation. As more and more people all over the world consume more food, drive more cars, use more manufactured goods, and use more fossil fuels, then more heat-trapping carbon dioxide builds up in the atmosphere.

So, back to the radio talk show. The hosts' sarcastic questions deflected serious consideration of the significant problems associated with population growth. Yes, they conceded, life was a little more complicated and traffic considerably more congested in the greater Denver metropolitan area than they were a few decades earlier. On the other hand, there were also better professional sports teams and more cultural happenings of all sorts.

Colorado and its neighboring states still have space to accommodate more people if you don't mind urban sprawl, but water is the real catch. To their credit, the radio hosts then interviewed a city official in charge of Denver's municipal water supply, who was worried about being able to deliver enough water to drink and to wash the growing number of cars as the regional population continued to grow. Yes, the hosts agreed that water could be a problem for the American West, but engineers could probably find and deliver more water from somewhere, and besides, it will never stop raining. Some 15 or more years later, they might not be so cavalier in their assessment of the water problem, as the decades-long megadrought is having devastating effects on Colorado and the American West.

On one level, the answer to the question posed by the title of this chapter is simple: Increasing human population contributes immensely to our environmental and socio-economic problems. The municipal water manager got it right: more people mean more demand for water, which has been getting ever more difficult to provide in an arid climate like the American West. Even in humid climates, such as Atlanta, Georgia, population growth has outpaced the water supply, resulting in severe shortages and costly fixes.

While the talk show hosts were correct that it's unlikely to stop raining in Colorado, ongoing climate change has already reduced precipitation, reduced snowpack, and increased the length and variability of droughts in the American West. We were unaware back in 2006 that the West was only a few years into what has turned out to be more than a two-decades-long event, exacerbated by climate change and comparable to historical megadroughts.[3] Moreover, while the term *drought* implies a temporary situation that will eventually run its course, the current climate change in the region is more aptly called *aridification*. In addition to a long-term trend of less precipitation (drought), the American West is also getting hotter due to climate change. Of the precipitation that does fall, more of it evaporates under the hotter conditions and less runs off into rivers. Taken together, the drought and the heat are causing a more arid climate and less river flow.[4] The giant reservoirs of Lake Mead and Lake Powell hit record low water levels in 2021.

As these climate change trends continue, the water manager's job will become much, much more difficult, because river flows are decreasing and demand for water is outpacing supply. The engineering solution—bringing in water from some other region—only robs Petra to pay Paula, as the population and the demand for water is also growing in Petra's neck of the woods, too. The environmental and economic consequences of moving large volumes of water long distances are not trivial. Regional and local water shortages have reached the point where some communities are beginning to prohibit new growth.[5] This challenge in the American West is being played out throughout the world, such as water shortage emergencies in Cape Town, South

Africa, and in Chennai, India, where municipal water reservoirs have come dangerously close to drying up altogether.[6]

As with the challenge of supplying water, it will also be immensely challenging to feed more people without cutting down more tropical forests to create farms and ranches, using up groundwater for irrigation, and polluting rivers, lakes, and estuaries with fertilizer runoff. In the Punjab, the "food bowl of India," which has helped that country become mostly self-sufficient in food production despite rapid population growth, one only needed to dig about 3 meters (10 feet) to find well water in the 1960s, but today new wells must be dug as deep as 150 meters (about 500 feet). The cumulative effect of decades of irrigation in the Punjab have exceeded groundwater recharge from rainfall, especially during the last ten years.[7] The examples of challenges resulting from growing population could go on and on, such as the sea of plastic accumulating in the ocean because so many more consumers are using plastic bags and other plastic items once and throwing them away (discussed in chapter 7), but I'll stop here for now. Suffice it to say that these myriad problems would be easier to solve if there were fewer people consuming resources and producing waste.

How Many Babies Will Our Babies Have?

The girls are already alive who will become the mothers of most of the babies born 15–30 years from now, and so, depending upon the average number of children born per woman, demographers can make pretty good projections about the likely range of population growth that will occur between now and 2050 as those young girls become women of reproductive age. Beyond that, the projections are less certain. The good news is that the rate of population growth is slowing because families are having fewer children on average, and so the global population will probably stabilize by about the year 2100. The bad news is that the population

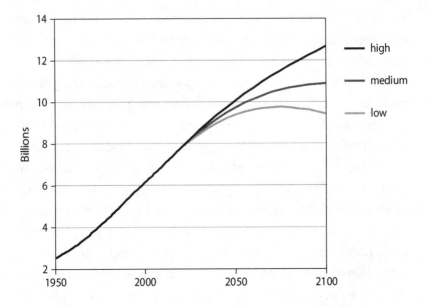

FIGURE 3.1 Projections of Population Growth. World human population is plotted from 1950 to the present and then projected to 2100 with high, medium, and low probabilistic population scenarios. Redrawn by the author from United Nations, Department of Economic and Social Affairs, Population Division, *World Population Prospects 2019*, online edition, rev. 1, https://population.un.org/wpp/

will probably not stabilize until it reaches at least 9.5 billion, and it could go as high as 12.5 billion before leveling off (fig. 3.1).

There is a big difference between 9.5 and 12.5 billion people projected for 2100 (the light gray and black lines in fig. 3.1), but that reflects the uncertainty of demographers' population estimates for the end of the century, when many of today's children will still be alive as senior citizens. What we do between now and then will determine where in this range the population will fall. Education and employment opportunities for girls and women outside the home and the availability of effective birth control are the most important factors affecting population growth across a broad range of countries, cultures, and religions.[8] The future global population is not predestined or inevitable; rather it will depend upon our current actions and will have consequences that last for generations. The needed response is not only about environmental policy per se, it must also permeate all social and economic policies, including promoting equitable opportunities for women and girls in the classroom and in the workplace. That said, we need to get ready for at least 9.5 billion people, and probably more, so we need to find a means to accommodate them—and not just accommodate, but enable them to prosper and lead meaningful lives.

A Dated View from the White Old Deal

The common neoclassical economic dogma suggests that human population growth is good for economic prosperity because it means more workers. That view is dated, as most economists now recognize that the factors explaining economic growth are very complex. Indeed, in *Good Economics for Hard Times*, Abhijit Banerjee and Esther Duflo, winners of the 2019 Nobel Prize in Economics, explain that economists have long used a jargon term— total factor productivity (TFP)—"to explain what is left after we have accounted for everything else we can measure" concerning the causes of economic growth.[9] In many cases, the TFP leftover is bigger than what can be attributed unequivocally to other known drivers of economic growth—in other words, economists don't really know what causes economic growth. Among the many difficult-to-quantify contributors to TFP growth are the skills and ingenuity of the workforce, not just the number. The novel ideas that people create for innovations and new technologies are essential for making more wealth with the natural and human resources at hand and in a manner that does not deplete or pollute those resources for future generations.

One concern about slowing population growth in many developed countries is that the population is growing older on average, so that fewer young people of working age will be supporting a larger population of retirees. That demographic shift has been clear for several decades now in Japan and Italy, for example.

It is also an issue in several other European countries and is even a new concern for China. How to support an aging population is a legitimate concern. The potentially destabilizing social and political implications of a smaller workforce and potentially underfunded social safety nets, like social security and Medicare in the United States, are real and should be addressed.

There are many good reasons to support family-friendly policies, but they are mostly independent of population growth. Granting mothers adequate maternity leave and assisting with costs of day care and after-school programs may increase fertility rates modestly, but the main justification for these policies is to allow women to pursue careers and motherhood as they choose. Enacting targeted tax deductions for dependents and funding bonds set aside for babies can help lower-income families provide an economic footing for their children. But tax deductions for children of higher-income families simply to encourage population growth of that segment of society are difficult to justify on their own.

If increasing the number of young workers is, in fact, desirable to support an otherwise aging population, an effective and rapid means would be to allow more immigration, which brings in mostly young workers and their children. Unfortunately, some politicians (usually white males) exhort Americans and Europeans to have more babies because there are too few workers (and they even boast about how they did their part by fathering several children). Such calls for higher birth rates in developed countries appear to be a thinly veiled appeal to keep those nations ethnically "pure." Indeed, some white supremacist groups state this goal overtly. Such racism is not unique to the West. Why does the Chinese government need to encourage higher birth rates among its majority Han ethnic group while discouraging its several Muslim ethnic minority groups from having children?

Racism aside, why would we need to encourage larger families in the United States and Europe when there are so many young, industrious, and clever workers in the world who would welcome a chance to migrate to a land of opportunities where they can prove themselves? Have immigrants not been a source of great innovation and increased productivity in America's past and throughout the world? Do we think of the families of Henry Ford, Steve Jobs, Sergey Brin, Colin Powell, Liz Claiborne, Arianna Huffington, and Jeff Bezos as Americans or as immigrants? They are both. Would America have been better off keeping them or their parents away? My own great-grandparents and grandfather immigrated through Ellis Island in 1901, which means that my parents, siblings, cousins, and I are also cut from the same immigrant cloth.

Current immigration is widely challenged by the misconception that the millions of immigrants who come to the United States and Europe are taking away jobs from local citizens. In fact, most immigrants fill jobs that few locals are willing to do, such as cleaning houses and hotel rooms, harvesting crops by hand, butchering animals in meat packing plants, and providing low-paying elder care and child care. But let's not rely on hearsay—what do the data say? Banerjee and Duflo provide data demonstrating that, in a few instances, immigrants may take a job that a local citizen might have applied for, but in most instances, they fill unwanted jobs. They then spend much of their earnings at locally owned businesses, which creates more, not fewer, jobs for American and European citizens.[10] They also pay taxes whether they receive benefits or not.

Banerjee and Duflo's evidence might seem to fly in the face of classic supply-and-demand theory, which predicts that an influx of immigrants will over-supply the labor force and thus draw down wages and job opportunities for the native residents. Supply-and-demand thinking makes some intuitive sense for labor, and thus politicians often convince voters that it is true, but economists have repeatedly demonstrated that it seldom works out that way for immigration. Again, Banerjee and Duflo document how study after study in countries throughout the world have shown no long-term negative impact of immigration on wages of local citizens. Many show net job creation and increases in wages because the immigrants' presence stimulates the local economy as they spend the wages earned from the jobs that few others want. At the end of the day, politicians striking up fears of job losses due to waves of immigrants are usually appealing to racist xenophobia without any real evidence that employment would be harmed.

As immigrant children become educated, they can make further contributions to workforce strength and innovation beyond the often menial jobs that their parents had no choice but to take. In that sense, my family's experience is again relevant—my great-grandfather, although a factory foreman in Scotland, worked as a janitor after arriving in America, which was probably a job that no one else in Great Falls, Montana, particularly wanted in 1901. His descendants, however, became teachers, artists, professionals of various types, and small and large business owners. It is important to add the caveat that the varying degrees of opportunities and wealth that my family members accumulated along the way were assisted by the many privileges that came with becoming fully assimilated descendants of white European immigrants. However, many immigrant families of color tell similar stories, despite the additional barriers that they encountered. Indeed, take the example of the generation of immigrants who are now called "Dreamers" because they were brought illegally to the United States about twenty years ago as children accompanying their parents. They have since largely assimilated into American culture to the point where they know no other home. Many have become young professionals, and some are frontline health care workers who have been risking their lives to help care for other Americans who have fallen victim to the pandemic. What larger contribution to America could we ask of immigrants? These immigrants are also among the young adults who are paying into social security, helping keep that system afloat to serve an aging baby boomer population of retirees.

While we have a duty to call out racism regarding population growth and immigration, a long-term objective should encompass understanding *why* people feel genuinely threatened by immigrants. In addition to the jobs issue, for example, immigrants are often perceived to generally disrupt "the way things have been." In this regard, the needed convergence across disciplines does not stop with the environmental and economic implications of population growth and resource depletion, but also includes exploring historical and socio-economic reasons for racial and ethnic tensions. Policies promoting segregation by ethnic identities reinforce distrust and fear of the opposing unfamiliar group, whereas those fears can be broken down with repeated, respectful interactions promoted by policies of inclusion.

Environmental Refugees

Living with more people on Earth is a given, but how many more, where will they live, and how they will flourish? The 2021 Global Report on Food Crises found that 155 million people in 55 countries in 2020 were living in a phase 3 "crisis or worse" category of food security, defined as "high or above-usual acute malnutrition" or "marginally able to meet minimum food needs but only by depleting essential livelihood assets or through crisis-coping strategies."[11] The report indicates that the dominant causes for falling into this food security crisis category were conflicts (99 million people), economic shocks (40 million people), and weather extremes (16 million people). The coronavirus pandemic was included as one of the economic shocks. In many cases, however, these drivers are occurring simultaneously in the same location, leading some experts to use the term "syndemic" to refer to multiple co-occurring global scale crises having synergistic negative effects on human well-being.[12]

The numbers of people displaced by conflict, natural disasters, such as floods and hurricanes, and longer-term environmental stressors, such as prolonged heat and drought, are already increasing as climate change and related syndemic conditions become more severe in the nations least economically capable of adapting.[13] The lead-up to the civil unrest of the Arab Spring, circa 2011, has been studied as an example of the important role of drought from climate change as a contributing stressor to civil unrest and conflict. Climatic stressors have also been linked to migration and an increase in asylum seeking.[14] The climate-conflict-migration linkage, to which a pandemic has now been added, is likely to grow with increasing poverty and civil unrest as lower-income countries are unable to provide resources to help their citizens adapt to climate change. The Red Cross estimates that over 30 million people were internally displaced by disasters in 2020, nearly all due to weather and climate hazards.[15] The World Bank estimates—conservatively they say—that over 200 million people will migrate within their countries by 2050 due to the impacts of present trends of climate change.[16]

We Know How to Slow Population Growth, but...

We know what policies are needed to enable women and girls to have meaningful opportunities outside the home, which experience shows will also reduce average family size and projected future population increases. We know that global population growth will add stresses on our systems to provide people with adequate water, food, shelter, and energy at the same time that there are concerns about having enough young workers to support the aging cohort. We know that immigration can be managed in a way not to threaten jobs and even to enhance economic productivity. We have seen how successful policies of the mid-twentieth century narrowed income inequalities, lifted the lower and middle classes, and generated economic growth. We know that African Americans and other minoritized groups have not only been left out of most of that growth, but also remain more vulnerable to pandemics, pollution, and economic downturns. We have expertise in the natural sciences, engineering, economics, and social sciences to do better than just muddling through the Anthropocene. *But,* we have not yet tried to bring all of that knowledge and experience together into integrated approaches to take us beyond the muddling. Therefore, a green new deal is a welcome, plausible first step toward the needed integration.

In the following chapters, we will consider our current knowledge base for finding ways to grow the needed nutritious food, meet the energy and transportation needs without dangerously altering the Earth's climate, and create meaningful and just employment and economic opportunities for all of our present and growing global population. There are gaps in that knowledge to be sure, but there is more than enough to offer direction and hope and to motivate getting started.

If you were expecting a book about protecting the environment for its own sake, please consider expanding your horizons. Wilderness areas, national parks, and indigenous peoples' lands are incredibly important for ensuring habitat for a diversity of plants and animals and to ensure that there are still places of beauty and solace with minimal human influence. These are worthy topics, but I will address them here only in the context of how regenerative agriculture in combination with strong governance can reduce some of the pressures to clear more forests to create still more agricultural land. Likewise, the science of ecosystems now includes study of the pervasive roles of humans as part of every ecosystem in the Anthropocene, thus converging the science of sustainable, human-dominated ecosystems with the experiences, native knowledge, values, and ambitions of individuals, communities, and cultures.

The Earth will survive regardless of what we humans do to it, so a green new deal or conservation efforts under any other name are not really about "saving the planet," although that is a nice, catchy phrase. More accurately, we must steward the planet and its life support systems in ways that will allow us and a broad variety of our fellow Earthling plant and animal species to prosper in perpetuity. The environmentalism of a green new deal is about integrating and mainstreaming environmental concerns into everything we humans do.

Recommendations

The Green New Deal (H. Res. 109, 116th Cong., 1st Sess. [introduced February 7, 2019]) cites the many problems that are exacerbated by rapid growth of human population, such as climate change, access to clean air and water, hunger, nutrition, health care, and education, but it is silent on what to do about population growth and immigration. Women and migrant communities are among those included in its definition of "frontline and vulnerable communities" that have suffered from "systemic racial, regional, social, environmental, and economic injustices" and who should be a focus for "providing resources, training, and high-quality education, including higher education." While population is understated in the GND resolution, I would propose that the following somewhat more specific recommendations are entirely in the spirit of a green new deal:

Further promote education and employment opportunities outside the home for women and girls throughout the world, which are widely recognized as the most effective means of fostering demographic shifts toward smaller family sizes and which are key to stabilizing human populations. Regardless of population concerns, it is the right thing to do.

Develop new, affordable, and socially acceptable contraception technologies for men and women. This topic has gained renewed attention in recent years because new technologies in medical science research have opened new prospects for much more effective and convenient forms of contraception for women and men.[17] Of the 40% of pregnancies that are unintended globally, 24% occur because contraception was not used and 16% because it was used improperly or it failed.[18] These data suggest huge opportunities and need for improvement. To be successful, however, new products will require knowledge from medical science, economics, and social science to understand what technologies women and men are likely to adopt.[19] A green new deal should not only make existing products more accessible globally and support more R&D on new contraception technologies, it should also include policies to reverse the disincentives that have prevented pharmaceutical companies from developing and marketing new contraception products during the last few decades, such as tight regulations, high litigation risks, market uncertainties, and funding gaps.

Dealing with the causes of emigration at their source, such as political unrest, lack of human rights, organized crime, lack of employment opportunities, and environmental stresses of flooding, drought, and heat waves, will be the long-term solution to stem the trend of increasing migration pressure. We cannot pretend that these migration pressures will simply go away on their own or that they can be held at bay indefinitely with walls, guns, and ever-growing refugee camps. Such profound changes require recognition that the well-being of nations and societies are interlinked, so that a country's strategic investments in foreign aid outside its borders can be an act of self-interest as well as one of charity. For those immigrants already in place, *pathways to legal residence are needed so that immigrants' work and positive contributions to local economies (including tax revenues to keep social security solvent), as well as their human rights and basic needs, can be recognized and honored.*

4

Manure Happens

The Consequences of
Feeding Over Seven Billion Human Omnivores

When I was a kid, chickens didn't have fingers. Nor did buffalos have wings. When we went to McDonalds or the A&W Root Beer drive-in, we ordered burgers. Chicken came as drumsticks and breasts fried by a colonel from Kentucky. Since that time, happy meals with chicken fingers and cows promoting Chick-fil-A sandwiches on billboards and television have been brilliant marketing schemes. They help explain why Americans have gradually reduced their average per capita beef consumption by nearly half and than doubled their average per capita chicken consumption since 1975 (fig. 4.1). In addition to clever marketing, the poultry industry benefited from technological developments that enabled modern chicken and turkey factory houses to produce large volumes at relatively cheap prices compared with beef, albeit with considerable regional environmental costs.

Annual consumption of meat and fish in the United States peaked at about 237 lbs/person in 2004 (108 kg/person), declined slightly until 2014, and has picked up a bit since. The decline started a few years before the great recession of 2008, probably

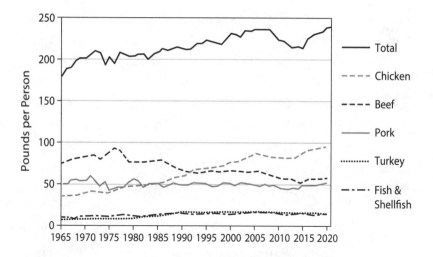

FIGURE 4.1 Per Capita Meat and Fish Consumption in the United States. Redrawn by the author based on data from the National Chicken Council, https://www .nationalchickencouncil.org/about-the-industry/statistics/per-capita -consumption-of-poultry-and-livestock-1965-to-estimated-2012-in-pounds/

due to higher commodity prices, and the recession probably contributed to a further decrease in meat consumption until 2014.

The average weight of adult American men is 197 pounds; for women it is 171 pounds (89 and 78 kilograms, respectively). This means that the average portly American consumes more than their body weight in meat annually (not counting animal products like milk, cheese, and eggs; the average amount of fish is relatively trivial). To fulfill this demand, we raised and slaughtered about 34 million head of cattle, 125 million pigs, and 9.2 billion chickens in the United States in 2018.[1] In China, with more than three times as many people and a strong preference for pork, the annual slaughter was about 40 million cattle, 700 million pigs, and 10.5 billion chickens. (China also has a live wild animal meat market, which is tiny in comparison to the numbers of livestock animals, but which may have profoundly changed the world as the possible locale of transmission of COVID-19 to humans, although the scientific evidence is not conclusive as of this writing.)

Globally, the 2018 slaughter was about 300 million cattle, 1.5 billion pigs, and 69 billion chickens.

With these huge livestock herds, manure happens in a big way. The global total manure production for livestock in 2018 was nearly 4 billion metric tons dry weight,[2] which is equal to the weight of the water in about 1.5 million Olympic-sized swimming pools. The wet weight of the manure would be five to ten times as much. Manure happens, literally and figuratively, with commensurately huge consequences for the environment and human health.

That much manure is happening now with "only" about 7.7 billion people on the planet, but barring a thermonuclear war or other catastrophes much worse than COVID-19, the human population will grow to at least 9.5 billion and perhaps as high as 12.5 billion by 2100 (see fig. 3.1). The magnitude of the challenges to provision that many people with food, water, clothing, energy, transportation, education, and the wherewithal to lead meaningful lives, without dangerously polluting the planet with excess wastes, will require the best convergent science and innovation. That much is clear, *but* we do not yet know just how we will produce the most essential of human provisions—food—and how it can be grown, transported, and consumed sustainably for 9–12 billion people. Therefore, we should heed the warning that "manure happens," both literally in terms of the enormous volumes to be reckoned with, as well as a figurative warning of the many environmental and human health perils awaiting if we do not intelligently and deliberately manage our food production systems.

Waste or Resource?

The growing amount of livestock manure poses challenges, but it's also a valuable resource. When applied to soils, it provides nitrogen, phosphorus, and other essential nutrients for crop growth as well as adding to soil organic matter. Unfortunately, only a modest fraction of manure from livestock is applied where it could do the most good. Instead, most corn grown in the US Midwest, for example, depends largely upon synthetic fertilizers, and only about half to two-thirds of the applied fertilizer makes it into the corn.[3] Most of the remainder gets washed into ditches and streams, is carried down the Mississippi River, and enters into the Gulf of Mexico, where it feeds noxious blooms of algae.[4] When the algae die, their decomposition by bacteria uses up the oxygen in the water, which kills off many fish in what has become known as "the dead zone." This zone of low-oxygen waters has extended over an average of about 5,700 square miles (about 15,000 square kilometers and a bit larger than the size of Connecticut) in the Gulf of Mexico each summer since the early 1990s.[5] Similar dead zones have emerged in estuaries and other coastal waters all over the world.[6] Despite losing about one-third or more of the applied fertilizer nitrogen and the resulting problems in the regions' water quality and in the Gulf of Mexico, agriculture in the US Midwest actually has more efficient use of fertilizer than most of the rest of the world, where losses of nitrogen to the environment are often even worse.[7]

Manure is often not used for its greatest value, which is to partially replace needed fertilizer on croplands. This is because many of the places where we raise animals for slaughter today, such as the large concentrated animal feeding operations (CAFOs) for hogs in North Carolina, are often far away from where we grow most of the crops for animal feed on vast areas of land, such as the corn belt of Iowa, Illinois, and Indiana. The manure is wet and heavy, so it is expensive to transport. Therefore, it is not economically valuable enough to cover the shipping costs from North Carolina all the way back to the cornfields of the Midwest, where it could substitute for some of the synthetic fertilizers to nourish the corn. Even shipping manure from hog operations in one part of Iowa to cornfields in another part of the same state may not be economical. In North Carolina, pig wastes are held in giant lagoons that often overflow into streams and rivers during storms. Depending upon how manure is stored, it can lose much of its beneficial nitrogen as gaseous emissions of ammonia and nitrous oxide, and it can produce a lot of the greenhouse gas methane. Eventually, it must be disposed of by spraying slurries of pig feces and urine with giant sprinklers on farms and even in forests of eastern North Carolina (fig. 4.2). Some of those forests, such as pine plantations, benefit from the manure spray, but this is not its highest-value use, and applications are often based on the need to get rid of the stuff rather than the ideal application rate for the trees. Ammonia gas emissions, pig slurry spray, and overflow into rivers and streams have become significant public nuisances.

This situation is the consequence of very deliberate policies. North Carolina provided financial incentives and lax environmental laws to attract the swine industry several decades ago when the demand for the region's tobacco crop was declining as efforts to convince Americans to stop smoking took hold.[8] Let's be clear, the buildup of the swine industry in North Carolina was not driven purely by free-market economics. Rather, it was the result of state government incentives.

North Carolina now has almost as many pigs as people, and its overall statewide economy has no doubt benefited from these policies, but the pig farmers and their rural neighbors? Not so much. The swine industry has adopted a trick also used by the poultry industry in other regions of the country. The pig and chicken farmers grow the animals from babies to adults under contract with the big slaughterhouses, but the farmers do not actually own the animals. The slaughterhouse corporations own the animals, and the contracted farmers just raise them for a fee. However, it is usually the farmers who own the manure that the animals produce, and it is their responsibility to dispose of it, not the responsibility of the big corporations that own the animals. As long as they have enough nearby farmland where they can use or sell this manure, it can be beneficial to the farmers, but the amount of manure usually exceeds the demand for nutrients on those nearby farms. A further complication is that pig and chicken manure is usually too rich in phosphorus relative to nitrogen, so that the balanced nutrient needs of the crops are not met. If enough manure is applied to meet the crop's need for nitrogen, for example, then there will be excess phosphorus in the applied manure that the crop cannot use, much of which runs off the field into waterways, where it also contributes to noxious algal blooms. Alternatively, if less manure is applied to avoid pollution from excess phosphorus, then the farmer must buy and apply supplemental nitrogen fertilizer to make up for the nitrogen deficit. Most importantly, the lack of sufficient nearby cropland on which to put the stuff means that farmers must still dispose of a large fraction of the manure somewhere else. The farmers' profit margins are slim, so they cannot afford to spend a lot on the manure disposal, and hence the widespread spraying of slurries of pig wastes onto pastures and forests (fig. 4.2).

Much of the nitrogen is also lost as ammonia gas from pig slurry holding ponds and from the sprayed slurry, which is a source of downwind pollution to waterways and estuaries. The slurry contains many other gases that are noxious nuisances and that present a serious public health hazard.[9] Besides the pig slurry's stench, the predominantly African American residents of the area cannot hang their laundry out to dry because droplets of pig slurry blown by the wind foul the clean laundry faster than it can dry.[10] Nor can they leave windows open or enjoy cookouts. Researchers have verified the source by identifying the pig DNA in the feces found on the sides of the houses. Although there have been some successful lawsuits, the state of North Carolina has modified its "right to farm" law to make it much more difficult

FIGURE 4.2 Spraying Pig Waste. A fixed spray gun irrigates a field with wastewater from a lagoon for storing hog waste in Warsaw, North Carolina. Photo by Donn Young, 2013

to hold accountable either the farmers or the slaughterhouse corporations that own the animals for how and where their manure happens.[11]

Carrots and Sticks for Farmers

What governments do with their incentives and policies to attract the swine industry or other industries, they could undo or modify to achieve multiple objectives of economic productivity, environmental protection, and environmental justice. Researchers are working on technologies to use the pig waste in anaerobic digesters to produce methane gas as fuel, and pilot plants are being launched using chicken manure.[12] However, a proposed biogas project enabled by a recent North Carolina farm bill is controversial because it would continue to rely on storing pig slurry in large lagoons and spraying it onto fields close to African

American rural communities.[13] Capturing the methane emissions from the lagoons of pig slurry to replace fossil fuel use elsewhere is a great idea, but that technology must be implemented with due consideration of environmental justice issues and engagement of local communities.

We should also provide incentives for farmers to reintegrate crop and animal production within the same areas, so that the animal waste is produced closer to areas where it can be effectively and economically recycled back onto cropland (we will return to the topic of a circular economy in chapter 7). Recycling the nutrients in the manure would also reduce the farmers' need for expensive synthetic fertilizers. Farmers know how to do this, and they would do it, given the right incentives. Instead, the swine industry in North Carolina and elsewhere created perverse incentives and consequences for the environment and for environmental justice. Appropriate incentives are needed that have positive impacts for the farmers' bottom lines, their neighbors' well-being, and environmental stewardship.

Many European governments take a regulatory approach, with fines for noncompliance serving as the farmers' incentives. I am not saying that is good or bad, but rather noting that it is different from the United States where the incentives in the agriculture sector are usually more carrot and less stick. Participation in the US is usually voluntary, and depending on the sweetness of the incentive, it can be effective. Maryland farmers, for example, routinely plant winter cover crops thanks to a state program that pays them to do so. The winter crop is usually a cereal like rye or wheat or a forage radish that grows well in cool weather and can survive temperatures slightly below freezing. They are not intended to be harvested for sale but rather are planted to help retain nutrients on the field during the winter season, suppress weeds, slow erosion, and improve soil structure, water infiltration, and aeration. Unfortunately, winter cover crops are still relatively uncommon throughout much of the American

Midwest, where such state subsidies are still rare. The 2017 USDA Agricultural Census reports that 4.8% of US croplands were planted in winter cover crops.[14] In Iowa it was only 1%–2%. The situation is beginning to change as innovative Midwestern farmers are demonstrating to their neighbors that they can save money on fertilizers and gradually build up organic matter in the soil by planting winter cover crops. Using winter cover crops can also be linked to reintegrating livestock with crop production because the livestock can graze on the winter cover crops, thus giving economic value to the cover crop as a partial replacement for costly animal feed while also recycling the livestock manure onto the cropland.

For either type of incentive, American carrots or European sticks, we must remember that decent economic returns for farmers are essential, and they can be compatible with environmental and social priorities if incentives are properly designed and implemented. Even after such enlightened practices are adopted, it takes some time for the soils to respond and for the benefits to be felt by the farmer and realized for the environment.

Interestingly, the American model also includes a significant role for private sector incentives, such as the major grocer Walmart. In a program codesigned with the Environmental Defense Fund,[15] Walmart requires the suppliers of some of its lines of groceries to confirm that the farmers who grew their produce and grain have nutrient management plans for the use of fertilizers, manures, and other nutrient inputs on their farms. Walmart claims to be doing this to provide its customers with environmentally responsible products. Of course, having a nutrient management plan, whether imposed by a retailer like Walmart or by a government agency, does not by itself ensure that every farmer carefully follows it or that the economic efficiencies and environmental benefits are fully realized. An important challenge for these innovative, incentive-based approaches is to establish reliable and cost-effective methods for accountability that the desired outcomes are achieved. Nevertheless, the partnership between a giant corporation and an environmental advocacy organization, agronomists, and farmers is a laudable first step. This approach is potentially a win-win-win proposition, in which consumers and retailers are pleased with the products offered, the farmers earn a premium price for participating, and the environment benefits from improved management that reduces nutrient losses to the air and water.[16] Together, they are searching for solutions, using science-based plans, with input and knowledge of farmers to manage the use of fertilizers and manure. General Mills, the food company that many Americans associate with breakfast cereal, announced a commitment to advance regenerative agriculture practices (described below) on one million acres of farmland by 2030, working with both organic and conventional farmers.[17]

Although an aside from the present focus on agriculture, let us not forget the complexity and interlinkages in our economy. For example, on the one hand Walmart is to be congratulated not only for its innovative efforts to require nutrient management plans on the farms that supply some of its grocery products, but also for increasing its production and use of renewable energy at many of its stores, for streamlining its shipping processes to conserve energy, for its leadership commitment to electrify and zero out emissions of all of its long-haul trucks by 2040, and for requiring all shoppers and employees to wear masks early in the COVID-19 pandemic. On the other hand, Walmart still carries countless nonrecyclable, single-use items and packaging in its stores that depend upon unsustainable resources and supply chains. Nor does it have a good record of accomplishment for paying all of its employees a living wage with decent benefits. Let the facts show where credit is due and where more accountability and progress are needed.

Principles of Regenerative Agriculture

The recent branding of several agronomic best practices under the umbrella of "regenerative agriculture" describes a holistic and systematic approach to increasing yields and profits while minimizing environmental impacts. Curiously, while regenerative agriculture is often touted as a recent revolutionary transformation, none of these practices is really new;[18] I learned about each while studying soil science as a graduate student in the 1980s. Moreover, there is no firm definition of what constitutes regenerative agriculture; it means different things to different researchers and practitioners.[19] Taken together and treated seriously, however, the following four principles emerging from this movement would advance sustainability in agriculture and further the goals articulated by many groups, from green-new-dealers to agricultural extension agents.

1. Minimize soil disturbance. The plow was a great invention that enabled farmers to harness the strength of horses, oxen, water buffalos, and eventually tractors to till the soil, overturning the unwanted weeds and creating furrows to plant seeds. The plow enabled farmers to produce more food per acre of land for their families and communities. Since then, however, agriculture has largely moved on to newer and better technologies in much of the world. You might still use your shovel, hoe, or rototiller in your home garden, but large-scale farming is increasingly using "no-till," which minimizes soil mixing ("no-till" is actually a bit of a misnomer, because most farmers who have adopted so-called no-till continue to till their soil once every few years).[20] Another practice is called "conservation tillage" which keeps much of the area covered with crop residues (for example, corn stalks or rice straw). By not mixing up the soil so much and by leaving crop residues on the field, organic matter can build up in the soil. Soil organic matter is one of those "mother-god-and-apple-pie" attributes of healthy soil that everyone agrees is virtuous, because it retains and gradually releases nutrients and water to the crops. Modern planting machines can easily drill seeds into the untilled soil to the desired depth, so plowing is usually no longer needed to create furrows for planting. Controlling weeds is another matter, and we will return to that topic later.

2. *Diversify the mix of crops grown.* Vast uniform fields of all-corn or all-soybeans, also called "monocultures," are common in the United States and much of the world, but that is actually a fairly recent phenomenon of the last 50 years or so. The fields growing a single crop used to be smaller, there was a larger variety of crops grown among the many smaller plots, and animal production was often integrated into the farm system. Now, modern machines that are fine-tuned to plant, fertilize, and harvest corn and soybeans on vast areas each year are highly efficient and have contributed to increased productivity and profitability, but those crops are increasingly becoming vulnerable to pests and weeds. This large-scale monoculture is pejoratively called "industrial agriculture" by its critics, and it has become the norm in developed countries and so will be challenging to displace.

In contrast, multi-cropping can reduce the need for pesticides and fertilizers and prevent soil erosion, which should, in theory at least, improve profitability and sustainability over the long term. Crop mixes come in a variety of forms, from simply interspersing rows of another crop or native prairie plants amid the rows of corn, to complex mixtures of trees and annual crops growing together, with several options of intermediate complexity. Rotating the crops from year to year or planting different varieties of the same crop also helps to keep pests that specialize on a single crop or a single crop genotype (variety) from becoming ensconced at the site. There are many considerations regarding the management of shading, nutrients, and water supply. Also, the logistics and mechanics of harvesting become more complicated as the mixture of crops is diversified, which can require both more know-how and more labor. Increasing labor demands due to diversifying crop mixes may require rethinking current policies regarding temporary farm worker visas and their working conditions. The optimal crop mixes will vary by soil type and climate, and a great deal of experimentation by farmers and agronomic scientists will be needed. Funding for on-farm research and development and for incentives that reduce financial risks of experimental multi-cropping will be essential.

3. *Keep the soil covered at all times, preferably with live plants present year-round.* Bare soil is vulnerable to erosion caused by wind blowing the soil away in dust storms or by rainwater washing it away and forming gullies. Mulches between the rows of crops protect the soils from erosion and hold in moisture needed by the crops. Grasses or other cover crops protect the soil under fruit and nut tree crops and in vineyards. Leaving crop residues such as leaves and corn stalks on the soil after harvesting annual crops also reduces soil erosion and conserves organic matter and nutrients.

Living cover crops not only cover the soil, as their name indicates, but they also have live roots in the soil, which soak up nutrients that would otherwise leach away into the groundwater or streams. The nutrients taken up by the winter cover crop are turned back into the soil in the spring when the main crop replaces the cover crop, thus reducing (but not entirely eliminating) the need for added fertilizers and manures. The roots also infuse organic matter into the soil and support healthy communities of soil microbes that help release nutrients to crops. Winter cover crops that are adapted to growing under cold conditions, including surviving under snow for a while, are planted after fall harvest, slowing soil erosion and soaking up nutrients well into the winter.

We already discussed the encouraging, but still frustratingly slow trend of increased planting of winter cover crops in the United States. In Iowa, cover crops are used in about 1–2 million acres, and the state is trying to increase cover crop planting to

14 million acres, which would be more than 50% of the corn and soybean cropland. The state has offered a $5 per acre discount on crop insurance for farmers who use cover crops.[21] A ten-fold increase in adoption of cover crops seems like a lot, but it is doable. There has been a similarly huge increase in adoption of conservation tillage in the United States and throughout much of the world. Adoption of cover crops is a bit more complicated, because one size certainly does not fit all, as the right combination of cover crops and management approaches vary with soil type, climate, and the rest of the cropping and animal production system. Like any crop, cover crops need informed management that integrates them into the whole farming system in order to provide their full potential benefits to the farmer and the environment. With the right incentives that use the convergent knowledge of farmers, crop advisors, soil scientists, social scientists, and economists, winter cover crops should become the norm throughout temperate regions. In tropical regions, more than one crop is often grown per year if there is enough rain, and drought-tolerant cover crops can protect the soil where there is a significant tropical dry season.

4. *Reintegrate livestock with the crops.* Children learn the sounds made by farmyard animals by singing songs like *Old McDonald Had a Farm*, but sadly, children would not be able to find a moo-moo here and an oink-oink there on most large-scale US farms today. With some important exceptions, most of the livestock are located in specialized large livestock facilities at great distance from the farms that grow their feed.[22] Tragically, many of these animals are raised under crowded conditions that you would not want your kids to see or sing about. We already discussed the wasted manure resource and the manure disposal problems caused by the example of separating grain production in the US Midwest from hog production in North Carolina. When crops and livestock are produced on the same or nearby farms, the expense of transporting the manure is more reasonable, so that

the manure becomes an economically valuable asset as well as a soil builder and an effective substitute for some of the fertilizer needs. Mobile chicken coops rotated around a fallow field allow the chickens to feed on insect pests while enriching the soil with their poo-poo here and their poo-poo there. Likewise, rotating cattle around a pasture keeps the cattle from eating too much of the grass and actually stimulates the remaining grass to grow better after the cattle move on to the next area. Putting cattle on the cropland during winter can also provide a good economic advantage for cover crops as the cattle graze on the cover crop, gain weight, and drop their manure on the site. Smart ranchers have always rotated their cattle, but regenerative agriculture requires them to pay closer attention to how often and where the rotations should occur. This system can also be integrated with the principle of maximizing crop diversity by alternating row crops and pasture for livestock every few years.

One of the barriers to reintegration of livestock into cropping systems is that many of the current generation of crop farmers have never raised livestock before (Old MacDonald had his farm several generations ago). So even if bringing animals back onto the family farms that they inherited may make good sense, it would require learning new farming practices and taking on significant new financial risks. With appropriate incentives and agricultural extension outreach, that should not be an insurmountable barrier, but it illustrates that farming depends upon human resources.

How Productive Agriculture Can Help Conserve Forests

The fates of forests and food production are intertwined. Conservationists who study what needs to be done to slow or stop the extinctions of species of plants and animals in forests and native grasslands around the world have concluded that it cannot be accomplished without concurrent innovation in food production systems and human diets so that more nutritious food can be grown for affordable prices on less land.[23] At first blush, the logic for why yields on existing cropland must improve in order to avoid cutting down more forest is simple:

- A growing human population needs more food (even after taking into account reducing food wastage and possible changes in diets, which will be addressed later in this chapter).
- More food requires either (A) greater crop yields on existing farmland or (B) more farmland.
- If we have to make more farmland under option B, that will come from clearing forests and woodlands, which would reduce habitat and other services those ecosystems provide to wildlife and to humanity.
- So, if we want to conserve existing forests, we need to choose option A to increase yields on existing farmland.[24]

A caveat to that logic, however, is that even with greater crop yields on existing farmland, there will still be pressure to cut forests to make money from productive agriculture, and so there is also a parallel need for laws and incentives for nations to conserve their forests, savannas, and grasslands. A distinguished group of conservation scientists have mapped out the regions of the world where land could feasibly be conserved to meet the goals of protecting 30% of the land area under its native vegetation by 2030, which they call the Global Deal for Nature.[25] They present compelling arguments for conserving biodiversity of plant and animal species, protecting clean water sources, and sequestering carbon from the atmosphere. However, initiatives to protect areas of native vegetation also need a convergence with the social sciences to understand how local communities of indigenous people will be affected.[26] Conservation does not necessarily mean excluding people. In fact, areas occupied by indigenous people are often more effectively protected than are national parks. There are similar proposals for protecting large areas of the ocean.[27]

This is unlikely to happen, however, unless we can increase food production on existing cleared land and do so in a manner that can be sustained for generations—that is, regenerative agriculture. Raising the standard of living and the nutritional well-being of the people of developing nations will increase pressures to clear more land for agriculture, which (if allowed to follow historical patterns) will have the dual consequences of increasing heat-trapping greenhouse gas emissions that cause climate change and reducing habitat for plant and animal diversity. The only way out of this dilemma is developing and adopting more efficient, productive, and sustainable technologies and management practices in agriculture, accompanied by governance structures that limit areal expansion of agricultural lands.

Yet one more frightening concern about deforestation for agriculture is emerging in this pandemic-conscious world. Climate change and continued expansion of human populations into patchworks of recently deforested areas may enable more transfers of diseases from wildlife to humans. Large animals with highly specific feeding or habitat requirements, such as big cats, rhinoceroses, or ostriches, tend to die out when their habitat is largely deforested, but many smaller and more abundant animals, such as rodents, bats, and some small birds, can partially adapt to a landscape with intermittent forest and cropland. It turns out that these same little survivors are also likely to host pathogens that can jump from the animals to infect humans.[28] COVID-19 could be followed by COVID-29 or COVID-xx, along with other zoonotic diseases (that is, diseases transmissible from animals to humans under natural conditions), unless we find ways to keep forests intact and productive agriculture in its proper place.[29] Investments in programs to reduce tropical deforestation are likely to pay for themselves several times over when compared with the costs of a global pandemic.[30]

But let's not stop there. Indeed, we need to go beyond conserving existing forests; we also need to reforest lands that were only marginal for agricultural productivity to begin with and that have since become degraded. By allowing forests to grow back on marginally productive agricultural land, carbon will be withdrawn from the atmosphere and stored in the woody biomass of trees and in the soils. At the same time, habitat for diverse species of plants and animals would also be restored. Forest regrowth alone will not be enough to counteract emissions of CO_2 from fossil fuel combustion, but it can be an important wedge in the multipronged strategies to mitigate climate change and increase native habitats for diverse species of plants and animals.

Expanding aquaculture provides another opportunity for meeting growing global food demands. Although fish and shellfish contribute only a trivial amount of protein to the average American diet (see fig. 4.1), fish accounted for about 17% of total animal protein consumed by humans globally in 2017.[31] Nearly 50% of that fish comes from aquaculture, which has been increasing while the wild fish catch has stagnated as many wild fisheries are becoming depleted. The trend of increasing aquaculture production offers opportunities to improve human nutrition, contribute to local economic productivity, and reduce demand for clearing more cropland. However, aquaculture has its own set of challenges for sustainability. These include pathogens, parasites, pests, waste disposal, nitrogen and phosphorus pollution, land and water needs, greenhouse gas emissions, and finding sustainable sources of fishmeal, all of which often become problematic as the technology intensifies and scales up to produce larger quantities of fish and shellfish.[32] In other words manure happens, literally and figuratively, for fish, too.

Incentives for Regenerative Agriculture

I do not believe that most aspects of regenerative agriculture can or should be forced upon farmers by national legislation, because the best regenerative agriculture practices are likely to be locally specific. While we know the basic principles of regenerative agriculture, the devil—or the god, depending on your point of view—is in the details. In other words, regenerative agriculture depends upon a lot of knowledge and experience from multiple and often local sources. It will require trust among farmers, extension agents, private sector consultants, supply chain corporations, and scientists to develop the optimal details of regenerative agriculture for each region and community. National and state policies could certainly encourage these efforts with enlightened financial incentives and support for both basic and on-farm research that includes agronomy, economics, and social science.

Government policies alone cannot create trust, but they can enable it by incentivizing the parties to work together in search of solutions. Innovative agricultural research and extension must engage farmers and crop advisors with agronomic scientists in co-learning efforts. Farmers will probably need appropriate financial incentives, such as properly targeted payments or insurance, to allow them to experiment with new management practices without taking on too much financial risk for their families. There can be an important role for some regulation, such as eliminating fall application of fertilizers in sensitive areas[33] and banning the more dangerous pesticides (discussed below), but we must also acknowledge that an onerous governmental regulatory approach is generally not a trust-builder, especially not in American agricultural communities.

Not Conflating Regenerative Agriculture and Organic Agriculture

By now, you may have gathered that, while I am an enthusiast of innovative, knowledge-based, regenerative agriculture, I am not a hard-core proponent (nor am I an opponent) of organic farming. I have no argument with organic farmers or consumers who prefer organically grown produce, but I just don't see how we can feed 10-12 billion people without some judicious use of fertilizers, herbicides, and perhaps even some pesticides. Most importantly, high-knowledge and sometimes high-tech approaches will be required.

Replacing our current "industrial agriculture" with regenerative agriculture practices will not necessarily do away with large fields and big machines in developed countries like the United States. The principles of regenerative agriculture are mostly independent of farm size. Indeed, aside from fewer examples of pure monoculture dominating vast landscapes, large-scale regenerative agriculture may still look rather "industrial" from the roadside as we drive by. Depending on hand-held hoes on small farms will not feed 10-12 billion people without cutting down more forests to create more small, low-productivity farms. American farms much less than about 500-1,000 acres (about 230-460 hectares) planted in staple crops cannot provide the economies of scale needed to reward farmers with profits approaching an American middle-class income unless they switch to growing relatively high-priced products, like berries, for niche markets. In contrast, farming on relatively small plots is still the dominant occupation in many developing countries, and most of those small farmers struggle to make a living beyond subsistence. Improving the livelihoods of small-holder farmers as some developing countries transition away from predominantly agrarian societies raises many complex socio-economic and environmental justice concerns. Each country must sustainably produce sufficient supplies of grains and other staples for local and global markets, using regenerative agriculture practices on farms of varying size, depending upon the country and its demographic structure and markets.

Adopting more plant-based diets and reducing food wastage (discussed below) would help reduce the demand for more grains, but production must still increase to nourish the growing global population.[34] Meeting that demand will require keeping nearly all of the tools in the toolbox, learning how to use those tools intelligently and safely, and embracing ideas and technologies that simultaneously advance food production, economic return, and environmental protection. There are many examples, but let us return to the challenging question that arose when discussing conservation tillage—how do you get rid of weeds if you don't till the soil thoroughly?

One answer is to reconsider what is a weed. Unlike Shakespeare's rose, a weed by any other name may not smell or look so bad. A weed is not necessarily a botanical term, although botanists often characterize weeds as species that tend to grow rapidly in disturbed habitats. More germane to this discussion is that a weed can be defined as a plant that is growing where a person does not want it to grow. If you don't want dandelions growing in your yard, then they are weeds, and you may either wield a dandelion digger or poison them with an herbicide. But if you don't mind a few dandelions and the many other clovers, plantains, and other common yard invaders, you can be less aggressive about removing them, and you might enjoy watching the bees that are attracted to some of their flowers. Likewise, some vineyards and fruit and nut tree farms encourage a non-crop ground cover to grow, such as grass, which helps conserve soil and its nutrients and promotes a diversity of beneficial soil insects and microorganisms that offer protection against other pests and erosion.

On the other hand, if left unchecked, weeds tend to overgrow the desired crop plant, and knocking them back is often necessary. Some organic farmers, especially those farming small plots for local markets of customers looking for organically grown produce, successfully use mulches to smother the weeds, and they pull or hoe other weeds by hand or use a rototiller. That kind of farming requires a large source of mulch and a lot of human labor. The hoeing and tilling avoids the use of herbicides, but it also causes loss of soil organic matter from all of that soil mixing. Larger-scale farmers who use no-till or conservation tillage practices to help regenerate and maintain healthy amounts of soil organic matter usually use herbicides to control weeds in their no-till fields. Likewise, winter cover crops may need to be killed in the spring before a new crop can grow, and that is often done with herbicides. So sometimes the principles of regenerative agriculture, such as no-till and use of winter cover crops, part ways with strictly organic agriculture. A new tool called a roller-crimper has been developed to mechanically push down rye winter cover crops without herbicides, but the developers and farmers are still working out the bugs in terms of when and how to use it with consistent effectiveness.

In some cases, farmers plant genetically modified crops (GM crops) that are resistant to the herbicides used to control the weeds in the untilled soil. Can GM crops, herbicides, and pesticides be used judiciously and with minimal risk to farm workers, consumers, and the environment? To be sure, there have been some disasters that should never be repeated, such as the story of pigweed.

Throughout much of the United States and in many other countries, farmers have planted soybean varieties that have been genetically modified to be resistant to the common herbicide glyphosate, also known by its brand name, Round-Up. For several years, the GM soybeans grew great, while the weeds were very effectively controlled by the herbicide. After a while, however, a plant called *pigweed* that grows in the southern United States started to develop resistance to glyphosate. Evolution happened right under our eyes, as pigweed plants with natural mutations that rendered them resistant to glyphosate were able to survive and pass down those genes to their seeds, which sprouted and grew in greater abundance the following year, and the following year, and so on. Eventually, glyphosate-resistant pigweed overran vast fields of glyphosate-resistant GM soybeans in the southern US (fig. 4.3), causing soybean yields and farmers' profits to plunge. Other herbicides, like Dicamba, were tried, but pigweed developed resistance to them too, and the Dicamba also drifted in the air to neighboring farms, where it affected the neighbors' crops. In 2017, the International Agency for Research

FIGURE 4.3 Pigweed Overgrowing Soybeans. The taller, darker, oval-shaped leaves with emerging flowering tassels are pigweed; the more pointed, lighter leaves on shorter stalks are soybean. United Soybean Board. Obtained from Flickr under Creative Commons license, https://bit.ly/3jJro4S

on Cancer (IARC) classified glyphosate as "probably carcino-genic," indicating that more research on its impacts on human health is needed.[35] In the meantime, herbicide-resistant weeds have become so common that some farmers are starting to abandon no-till (which, as we saw earlier, helps build up beneficial soil organic matter) because they have no good option other than tillage to control the aggressive and abundant weeds.[36] In some cases, cover crops, which we discussed above with respect to nutrient retention and soil erosion, can have the additional benefit of helping control the emergence of pigweed.

The chemical and seed companies are already developing other herbicides and GM crops that will be resistant to them,[37] but we seem to be losing the race, as weeds develop resistance faster than new herbicides are developed.[38] Evolution is a strong force that enabled mosquitoes to become resistant to DDT and the malaria pathogen to become resistant to several antimalarial drugs, and it will enable many weeds to develop resistance to herbicides when used with abandon, as was the case with the glyphosate-pigweed story. This arms race between pests of all kinds and chemicals designed to control them was eloquently described over a half-century ago by biologist Rachel Carson in her book *Silent Spring*: "The chemical war is never won, and all life is caught in its violent crossfire."[39]

Rather than waging all-out chemical war against weeds and insect pests, we should keep our powder dry and use it sparingly only as one of several defense strategies. Herbicides and pesticides can be used properly and judiciously as part of a suite of carefully coordinated management actions called "integrated pest management" (IPM).[40] Knowledge about the life cycles of weeds and insect pests reveals all sorts of options, such as mixing in other species of plants with natural defenses, encouraging natural predators, and, yes, occasionally applying a dose of herbicide or pesticide at a key stage in the pest's development. Agronomists have advanced this knowledge-based IPM approach for decades, often with good success. Not every herbicide and pesticide is equal, and there are some that should be banned, such as the DDT that Rachel Carson studied, some organophosphates,[41] and some of the neonicotinoids that appear to be harmful to bees and other insects that pollinate crops.[42] No pesticide will be so safe that well-defined precautions are not needed or that excessive use can be tolerated, but that is also true of our medications, food, drink, and just about everything in life. Used with moderation, intelligence, and proper safety precautions, some herbicides and pesticides will be essential for feeding 10–12 billion people.

The emergence of swarming locusts in east Africa is a case in point, which is seriously threatening to cause famine throughout the region.[43] A new cell phone app that engaged local farmers to document the insects' spread allowed governments to target pesticide applications strategically to prevent further spread of this devastating pest. That use of pesticides, although targeted and limited, undoubtedly had some troublesome unintended consequences for other species of insects and for animals higher up in the food chain. Therefore, coordinated IPM efforts that control locusts before they begin to swarm would be preferable, so that the pesticide use can be further curtailed, but it may not be eliminated entirely. Some of my green new deal friends may disown me for my openness to a limited role for pesticides where they can be intelligently targeted. Nevertheless, I believe it must be evaluated on a case-by-case basis in the needed convergence of good agronomic, health, and social sciences to bring forth the best combination of options for human nutrition, safety, and the environment.

Judicious use of herbicides and pesticides is comparable to the judicious use of antibiotics, which revolutionized medicine and extended life expectancies, and which few of us would want to abandon. They, too, can be considered part of what Rachel Carson called a chemical war, albeit against human pathogens rather than agronomic pests. When antibiotics are recklessly overused, pathogens can and do quickly evolve resistance to them, and there are many previously effective drugs that are now essentially useless. The same is true for previously effective antimalarial prophylactic drugs. Antibiotic resistance is also growing because of widespread use at low doses to try to keep livestock healthy.[44] Developing resistance is probably inevitable in the arms race of evolution, but using antibiotics only under strict guidelines extends their useful lifetimes before resistance develops, giving us time to find other treatments. Even vaccines are part of the chemical war. At the time of this writing, the world is in a race against variants of the COVID-19 coronavirus to get people vaccinated before variants emerge that are resistant to the current generation of vaccines.

Given the rapidity with which pests, weeds, bacteria, and viruses evolve to avoid our cleverest efforts to keep them at bay, we are destined to remain in Carson's chemical war to some degree. Yet, I am optimistic that we can hold our fire when there are alternatives and that we can manage the crossfire intelligently when interventions are necessary to ensure the availability of food, medicine, and vaccines needed to provision and protect 10–12 billion people. An important difference between pesticides (including herbicides) and antibiotics, however, is that the latter are toxic only to a limited number of people with specific allergic reactions, whereas widespread use of pesticides can result both in the target pest developing resistance and in nontarget species, including humans, being harmed by the toxic pesticide.

Genetically modified crops pose a different set of opportunities and challenges. In my view, it was prudent to proceed very cautiously when GM organisms first emerged as a technological possibility, because the risks were largely unknown. However, much research since the 1990s has demonstrated that the fear of "Franken-foods" has been greatly overstated.[45] Appropriate scientific protocols for evaluating the safety and benefits of several

GM crops have since been established, although they must continue to be scrutinized, kept up to date with emerging technology, and diligently enforced.

An excellent example of the potentially huge benefits of a GM crop is Golden Rice. This crop has been modified to be extra rich in vitamin A. It has been approved in the Philippines and will likely be approved in Bangladesh. Vitamin A deficiency is the world's leading cause of childhood blindness, and 21% of Bangladeshi children have vitamin A deficiency.[46] Yes, they probably could get their vitamin A from sweet potatoes, carrots, papayas, and other A-rich fruits and vegetables or from vitamin-fortified imported foods, but for cultures in which rice is such a major part of the diet, Golden Rice may offer a culturally sensitive solution to a very serious malnutrition problem. The convergent science needed must integrate the safety and cultural concerns of consumers and farmworkers with the other environmental, economic, and health challenges inherent in nourishing the human population.

We Are What We Eat, and So Is Our Environmental Impact

I started this chapter revealing that I am too old to have grown up with chicken fingers and happy meals, but my dad loved chicken drumsticks. His idea of giving my mom a break from cooking was to bring home a bucket of fried chicken from KFC. I also remember that he had bacon for most breakfasts, cold cuts in his sandwiches at lunch, and either chicken or a hunk of red meat for dinner every evening. With that upbringing, it isn't hard for me to choose a diet now that is more environmentally conscious and healthier than was my dad's.

What are the environmental consequences of eating so much meat? We have already discussed the serious problems associated with disposing of or recycling the vast amounts of manure produced by livestock. Perhaps even more important is concern about how much cropland, irrigation water, and fertilizer we must devote to growing the crops for feeding livestock and the resulting emissions of greenhouse gases and leaching of pollutants into waterways. About one-third of all of the corn grown in the United States is fed to livestock, about half goes to biofuel, and only about 10% goes to other uses, including food directly for humans.[47] Much of that land, water, and fertilizer to grow all of that corn could be devoted to grains and other crops that humans could consume directly, which would feed more people and be more efficient. Livestock also graze on pasture grasses and consume feed that humans cannot consume, so not all of the livestock feed competes with human food production, but much of it does. Cattle and other ruminants also release large quantities of methane in their burps, which is a potent greenhouse gas. Pound for pound (kilo for kilo), growing cereals directly for human consumption is more effective at avoiding releases to the environment of greenhouse gases and water pollution by a factor of about three when compared with chicken and a factor of about ten compared with beef.[48]

One study has calculated that if current dietary trends continue, the contribution of the food production system to greenhouse gas emissions will nearly double by 2050.[49] However, if everyone adopted a "flexitarian" diet (defined as at least 500 grams per day of fruits and vegetables; at least 100 grams per day of plant-based protein such as legumes, soybeans and nuts; modest amounts of animal-based proteins, such as poultry, fish, milk and eggs; and one portion per week of red meat), the greenhouse gas emissions would not increase and instead would drop about 10% from present-day emissions. I am not necessarily advocating this specific diet, but the study's estimates demonstrate how important our dietary choices are. The same study estimates that cutting global food waste by half could contribute another decrease of 5%–10% in greenhouse gas emissions from food production. The entire food system, from the farms through the processing and supply chains, including packaging and transportation to consumers, and managing wastes, is estimated to be responsible for about one-third of all anthropogenic greenhouse gas emissions, so what we eat is nearly as important as the energy and industrial sectors as a source of global greenhouse gases.[50]

I confess that I'm not a vegan or vegetarian, but I watch the frequency and amount of meat that I eat, and I'm moving more and more toward a plant-based diet, both for personal and planetary health reasons. I enjoy a modest portion of a meat, poultry, or fish dish about two to four times a week and almost never more than once per day. I have to admit, however, that I am rather partial toward sharp cheeses. Mine is not a magic formula, and you may decide to eat more or less meat and dairy than me, with many options for delicious outcomes.[51] Meanwhile, we can all be cognizant of the impacts of our choices, both for our own health and for that of the planet. We all need easy access to good information about dietary choices so that each of us can make informed decisions based on our values and reliable information.

I remember that my junior high school cafeteria always served fish sticks on Fridays, even though Catholics were a small minority at my public school. Besides some grumbling about the tasteless frozen fish sticks, I don't recall any objection to a day without red meat. A few years ago, however, a US federal governmental agency started an innovative "meat-free Monday" initiative at its cafeteria to bring increased awareness to the environmental impacts of our dietary choices. When senators from states with large livestock industries got wind of this, they intervened to stop the initiative. If our culture tolerates pushing only fish sticks on Fridays, why not veggie dishes on Mondays? Fortunately, many colleges, universities, and private institutions are continuing to raise awareness about food choices in clever and innovative ways, including measuring their institution's "nitrogen footprint" and what can be done to reduce it.[52]

Contrary to some misconceptions and even purposeful misrepresentations, a green new deal would not force dietary changes on anyone or take away anyone's cheeseburger, but it would promote innovative efforts to encourage awareness of dietary choices for religious, personal, or planetary health reasons. Just as government warnings and funding of outreach efforts for kids and adults have played a key role in discouraging smoking without outright bans on all cigarettes, policies really can make a difference to encourage healthy choices among the vast array of food choices that wealthy societies are blessed to have and even among the more limited choices of less affluent communities. Of course, healthy food must also be readily available, which is not the case in "food deserts" of low-income neighbors without grocery stores—yet another challenge needing some green-new-deal convergent thinking. Not surprisingly, the issue is complex, requiring not only availability of markets, but also subsidies for healthy dietary choices.[53]

Less Food Waste Happens

It is difficult to talk about food waste without sounding preachy or moralizing. Indeed, my mom, like many mothers in her day, invoked "the starving children in India" as justification for why I should clean my dinner plate (which meant: "eat your vegetables, and don't let me catch you feeding them to the dog!"). Like many inquisitive children, I wondered how I could mail my Brussels sprouts to those hungry kids in India. FedEx had not yet been invented, so I couldn't figure out how to get those nasty things all the way to India before they would spoil (fig. 4.4).

My mother was right to admonish me to not waste food. About one-third of all food produced in the world is wasted—enough

When God created Brussels Sprouts

FIGURE 4.4 When God Created Brussels Sprouts. Out There by George, https://www.toonpool.com/cartoons/brussel%20sprouts_240604

to adequately nourish the 800 million or so people in the world who are currently malnourished. Since my childhood, India has become mostly self-sufficient in food production, although unequal distribution of food and wealth there (and throughout the world) means that many poor people remain malnourished. In the meantime, the groundwater in the Punjab has become depleted by the unsustainable irrigation needed to grow that much food, with the water table dropping about two and a half feet (75 cm) per year in some areas, rendering India's future food self-sufficiency at risk.[54] Meanwhile, sub-Saharan Africa has become the main geographic focus for fighting malnutrition. Regardless where it is—Asia, Africa, or the United States—both food wastage and malnourishment sadly and inexplicably coexist.

When I lived among mostly subsistence farmers in Africa, they could not rely on storing grains for either their own family's later consumption or for selling in a market, because rats and insects would eat it and mold would grow on it. Obviously, grain elevators and other effective storage technologies are common in developed countries, but they are too expensive for most African farmers. Technological progress is being made to develop inexpensive, simple, and effective small granaries, but they are not yet as widespread as needed. As a consequence, most of the wastage in developing countries occurs before food ever gets to market.[55] The technology exists, but the resources have not been allocated by governments to fix this very fixable problem.

In contrast, most food wastage in developed countries occurs post-market, such as those Brussels sprouts that my mom tried to get me to eat. If it were only a few vegetables left on the plates of finicky kids, it would not be so much of a problem, but about one-third of all food that goes to market in rich countries like the United States is wasted, a large part of it at the dinner table.[56]

Government policies should not attempt to replace our parents setting examples for their children's eating habits, but policy can have a big effect on several types of food wastage. It can provide incentives and mechanisms for grocers to transfer to food bank organizations the food that may be close to its "sell-by" date but that is still good. Government guidelines for food packaging currently confuse "sell-by" dates with the concept of an expiration date, causing many consumers (including my adult son, whom I cannot convince otherwise) to throw out perfectly good food that they think has "expired." It is important not to use medicines beyond their labeled expiration date, but "sell-by" dates for food have substantial built-in margins of safety for food to be taken home and consumed in the following days and weeks. The food is usually edible and healthy well after the sell-by date. Simple changes in labeling guidelines could avoid a lot of food wastage.

Those are just a few of many examples of why food is wasted. With proper incentives and resources, innovative thinkers in both private and public sectors could probably reduce our food wastage from its current shocking 30%–35% to about 10%–15%. I cannot think of a good way to encourage restaurants known for super-sizing their servings to match serving sizes with the varying appetites of their customers, but no doubt someone better informed or cleverer than I can come up with a workable solution for this and many other food wastage challenges. Significant reductions in food wastage would reduce the amount of fertilizer, pesticides, land clearing, and manure production caused by the current agricultural capacity to grow both the food that we eat and the food that we waste. While we cannot mandate such savings any more than my mom could always catch me sneaking Brussels sprouts to the family dog, we can support the social science and economic studies and the cooperation with the private sector needed to understand why people waste food and what can be done to effectively minimize it.

Habit Change Happens

Did I mention that Mom and Dad smoked, too? I was so accustomed to second-hand smoke that it seldom bothered me as a kid. Not so anymore. A profound, science-driven shift occurred within a single generation in our culture, making smoking in public places almost nonexistent in much of the world. Changes in smoking habits demonstrate that other human habits, from farming methods to dietary choices, can also change. The recent wave of vaping, popular among teenagers, shows that such battles may be never-ending, but we know that we can make a difference in helping people improve their own health and economic well-being, as well as sustain a healthy environment.

Lest we think that all changes must be made by farmers, corporations, and government policy makers, let us not forget that all of these crops and all of this meat is produced because we consumers are ready to buy it. Consumers could take more responsibility for being in the driver's seat based on our consumer demands. At the same time, we should not let corporations off the hook simply because they say that they are responding to consumer choices. Through their aggressive marketing and choice offerings, they have huge influences on consumers, for which they must be held accountable.

Farmers could grow a lot more crops for direct human consumption on existing land if consumers, either by their own innate preferences or through the influence of corporate marketing, did not demand so much meat and dairy. Similarly, less would have to be grown if we did not waste so much food. I would only grudgingly give up my sharp cheese, but, channeling Mom, we all have personal responsibility to make good dietary choices and to avoid food waste, thus helping to avoid shitloads (sorry, Mom) of manure happening on farms and its subsequent impacts on the environment.

BTW, Mom (may she RIP) would be happy to know that, I, too, have changed. Brussels sprouts are now my favorite vegetable.

Recommendations

The Green New Deal (H. Res. 109, 116th Cong., 1st Sess. [introduced February 7, 2019]) only briefly mentions agriculture:

> working collaboratively with farmers and ranchers in the United States to remove pollution and greenhouse gas emissions from the agricultural sector as much as is technologically feasible, including—
> (i) by supporting family farming;
> (ii) by investing in sustainable farming and land use practices that increase soil health; and
> (iii) by building a more sustainable food system that ensures universal access to healthy food.

Growing enough food on existing agricultural land will also be necessary in order to achieve the following stated GND goals:

removing greenhouse gases from the atmosphere and reducing pollution by restoring natural ecosystems through proven low-tech solutions
that increase soil carbon storage, such as land preservation and
afforestation;

restoring and protecting threatened, endangered, and fragile ecosystems through locally appropriate and science-based projects that
enhance biodiversity and support climate resiliency;

Although farmers are not specifically mentioned in the following list of stakeholders who must be engaged in co-production
of needed new knowledge, they certainly should be:

a Green New Deal must be developed through transparent and inclusive
consultation, collaboration, and partnership with frontline and vulnerable communities, labor unions, worker cooperatives, civil society
groups, academia, and businesses;

My guess is that the authors of H. Res. 109 shied away from
agriculture because of their limited experience and knowledge
on that topic. Clearly, more is needed, and the following recommendations for agriculture follow the same green new deal thinking as for other topics—namely, that the environment, economics, and social justice must be considered simultaneously through
transparent and inclusive initiatives.

Beyond the broad-brush GND goals for the United States articulated in H. Res. 109, agriculture in just about every country
in the world is strongly affected in multiple ways by its government's agricultural policies, often including many subsidies and
incentives. While the communist farm communes failed in the
former Soviet Union and under China's Great Leap Forward, we
should not fool ourselves that Western modern agriculture is entirely free-market driven. For example, about 40%–50% of the
corn grown in the United States goes to ethanol production,
which would be very unlikely in a free market situation, were it
not for federal subsidies and mandates.[57] Indeed, the US farm bill,

passed by Congress every five years, is always an enormous, complicated document filled with funding for all sorts of programs, incentives, and subsidies. Although not the only relevant legislation or policy instrument, the farm bill draws most of our attention in the United States, and many countries have similar comprehensive agriculture legislation. Here are a few recommendations that follow from green new deal thinking, that are more specific than the generalities of H. Res. 109, but still fairly high-level. The list is not comprehensive but focuses mostly on topics raised in this chapter:

Incentivize the adoption of regenerative agriculture practices, especially conservation tillage and no-till, diversification and rotation of crops, planting winter cover crops, and reintegration of livestock and crop production. Farmers will respond to incentives that reduce their financial risks while experimenting with novel practices. The Conservation Stewardship Program and other US Department of Agriculture conservation programs, for example, can and should be amended to increase the rewards and benefits of adopting regenerative agriculture practices. Expanded access to insurance products and preferential loans would help incentivize farmers to adopt management practices that might pose risks of not yielding short-term return on investments but that are known to provide longer-term benefits. A carbon bank is an idea to establish a fund that pays farmers for adopting best management practices that will increase soil organic matter. Validation is needed that soils actually and persistently sequester carbon over the long term, despite the common practice of occasional tillage under so-called no-till. Farmers will also need partnerships, which leads to the next recommendation.

Reinvigorate agricultural extension and on-farm research in ways that truly engage farmers and crop advisors to understand the many social, economic, and agronomic factors that influence their decision making. Declining state and federal budgets have decimated agricultural extension in the United States, and similar declines have been experienced throughout the world. Farmers need good information coming from the latest research, and farmers should also be welcome as meaningful partners with considerable experience and knowledge to contribute to designing and implementing on-farm research initiatives and co-producing new knowledge.

Support Integrative Pest Management practices and research and other thoughtful integrations of knowledge and technology to optimize crop productivity while intelligently minimizing uses, risks, and unintended consequences of herbicide, pesticide, and other pest-control technologies.

Encourage through regulations or strong financial incentives the adoption of effective nutrient management plans that are regionally appropriate for the crop and animal production systems, climates, soils, economies, and social contexts to reduce greenhouse gas emissions and reduce leaching of nutrients to groundwater and streams.

Eliminate subsidies for synthetic fertilizers in countries where they are often overused, such as China and India.

Transfer appropriate technologies and incorporate local knowledge to enable farmers in developing countries to increase yields on existing croplands while following regenerative agriculture practices. This includes ensuring that a variety of affordable nutrient sources are available, including fertilizers, manures, and mulches. Temporary subsidies for fertilizers can be justified in some developing countries, especially in sub-Saharan Africa, where they are currently unaffordable to most farmers, where soils have become severely depleted in essential plant nutrients, where alternatives such as manures are of low quality, where average crop yields are extremely low, and where malnutrition and food insecurity are very serious immediate problems.[58]

Make sure that agronomic initiatives are integrated with conservation policies for nearby forests, grasslands, and wetlands under native vegetation in support of the 30-by-30 initiative to conserve 30% of terrestrial ecosystems globally by 2030, consistent with adequate consideration of indigenous rights.

Phase out subsidies and mandates for corn-based ethanol. Although analyses in the last two decades indicate that increases in efficiencies of converting corn to ethanol have resulted in a net positive yield (that is, more energy in the ethanol product than the energy consumed to produce fertilizer and other products in the corn-ethanol production chain), this does not account for the other environmental impacts and the effects on food prices due to devoting large amounts of cropland to this form of biofuel production.[59] Consider phasing-in subsidies for perennial fuel crops, such as miscanthus and switchgrass, to be grown on marginal agricultural land with fewer energy and fertilizer inputs as their markets and appropriate technologies develop.[60]

Reduce food wastage by consumers and along the supply chain by harmonizing food labeling regulations to eliminate confusion about "sell-by" dates. Incentivize grocers, food supply chain companies, and restaurants to donate to food banks their edible food (or composting if inedible) that they can no longer use.

Reduce on-farm, post-harvest food wastes in developing countries by developing and disseminating affordable and effective grain storage technologies.

Make information on nutrition and dietary options more widely available to consumers and institutional planners through culturally sensitive and effective communication strategies.

5

Climate Change Viewed
by a Skeptic at Heart

Like most scientists, I am a skeptic. You could say that we are a
conservative bunch—not in the political left versus right sense—
but rather in that we hold on to old theories and widely accepted
explanations until the accumulated evidence becomes so compelling that a new scientific interpretation is necessary. Indeed,
that has been the case for the decades-long journey from initial
skepticism about the ability of humans to change the entire
Earth's climate to today's strong scientific consensus that human-
induced climate change is real, accelerating, and posing an existential threat to humanity.

From Healthy Skepticism to Scholarly Consensus:
A Journey of Scientific Integrity

Only 40 to 50 years ago, many scientists expressed a healthy
skepticism that humans could have a big enough impact on the
Earth's climate that it could be distinguished from natural climatic variation. I have witnessed the journey from that initial

skepticism to today's unequivocal scientific consensus that humans are changing the climate.[1] It is a remarkable story of the integrity of the scientific process, albeit with a historically interesting hiccup that predates the modern discussion.

The story begins more than a century ago. In 1859, John Tyndall, an Irish physicist, reported on measurements of the absorption of radiant energy by atmospheric gases. He is credited as the first to report the effects of atmospheric gas composition on the Earth's climate. A Swedish chemist, Svante Arrhenius, also made theoretical predictions in the 1890s, based on principles of physics, regarding the effects of atmospheric carbon dioxide (CO_2) on the Earth's climate. He also showed that CO_2 in the atmosphere has physical properties that amplify the warming effect of the sun by trapping heat within the Earth's atmosphere. Although he got the basic science right, he mused that it would be unlikely that humans would ever release as much CO_2 into the atmosphere as we subsequently did during the twentieth century.

Perhaps not too surprising for today's observers, a woman had actually beat Tyndall and Arrhenius to the punch, but until recently, her achievement has been mostly shrouded in obscurity.[2] In 1856, an American scientist, Eunice Foote, reported at a meeting of the American Association for the Advancement of Science on her series of experiments that demonstrated how the sun's interaction with atmospheric gases could affect the Earth's climate.[3] This is the same idea that we now call the "greenhouse effect" because CO_2 traps heat within the atmosphere much in the same way that the glass walls and ceiling of a greenhouse trap heat within the greenhouse on a sunny day. Her paper had to be presented on her behalf by a man, and it was left out of the meeting's official proceedings.[4] (Foote went on to be active in the movement for women's right to vote.)

Fast-forwarding 100 years, the first really good, consistent, and rigorous measurements of increasing atmospheric CO_2 began in the late 1950s as part of the International Geophysical Year, a coordinated effort by scientists around the world with support from their governments. In the 1960s, scientists started building mathematical models on newly emerging computers to help understand how varying amounts of atmospheric CO_2 and water vapor affect the global climate.[5] These models apply the same basic physics used by Foote, Tyndall, and Arrhenius. In addition to a warming trend documented in the early twentieth century, the start of a new and steeper warming trend was detected in the 1970s. Scientists then systematically set out to consider all possibilities, including, but not limited to Eunice Foote's greenhouse effect, for causes of these modern warming trends.

Scientists are trained to take a deliberative, stepwise approach to accepting new ideas. They posit hypotheses and then collect data that either support or refute those hypotheses, often testing and retesting hypotheses in new ways over many years. For example, testing a hypothesis that the late twentieth-century warming trend might have been caused by the variable strength of the sun, solar scientists concluded that the sun's strength and its sunspot cycles modestly vary the climate from one decade to the next, but the effect is far too small and too cyclical (warming and then cooling every eleven years or so) to account for the consistent warming trend observed over multiple decades.

Similarly, other skeptical scientists hypothesized that the asphalt, concrete, and steel being added to expanding cities were causing a heat-island effect. Perhaps, this hypothesis posited, weather stations that had once been outside cities had been affected by the heat islands that had overtaken them as the cities expanded. While heat-island effects are real, the hypothesis that they were responsible for the measured global warming trend was disproved. To be safe, surface temperature is now also measured on buoys in remote oceanic locations to make sure that records of temperature trends are unaffected by the heat from growing cities.

Volcanoes were also rejected as long-term warming agents because careful analysis showed that the amount of CO_2 emitted from currently active volcanoes is very small relative to the massive amount of CO_2 released from burning fossil fuels. Actually, volcanoes can temporarily cool the Earth's climate by ejecting huge quantities of ash and other particles into the atmosphere that reflect the sun's energy back out into space. Depending on the size of the volcano, it takes a year or more after the eruption for those particles to be washed out of the atmosphere by rain and for the cooling effect to go away.

On another line of inquiry about warming, subtle changes in the Earth's orbit around the sun have been shown to provoke the coming and going of ice ages. However, those changes happen over tens of thousands to hundreds of thousands of years, which is too slow to help explain the current rapid rate of observed warming over only decades.

Marcia McNutt, president of the National Academy of Sciences, and I summarized the history of investigating climate change in an article published by the American Geophysical Union:[6]

> By the end of the 1990s, these alternate hypotheses and others were carefully ruled out, one by one. . . . At the same time, a multitude of climate change trends became clearer, including higher surface temperatures and heat waves, melting Arctic sea ice, receding glaciers, rising sea level, changing patterns of extreme weather events, altered bird migrations, freeze and thaw dates of lakes, and so on. . . . The scientific community has gradually shifted, on the basis of evidence, from predominantly being skeptical in the 1970s that the human fingerprint on climate could be demonstrated to today being convinced that there are no other plausible explanations besides the cumulative effect of the last 150 years of burning fossil fuel for the recent extent of changing climate.

Our confidence is based on multiple lines of evidence: (1) it is consistent with well-known physics that was understood by Foote, Tyndall, and Arrhenius in the nineteenth century; (2) details of the patterns of climate change, such as temperature differences between the poles and the equator and between the stratosphere and the lower atmosphere, point to a human cause, much like a fingerprint identifies a criminal;[7] (3) the increasingly sophisticated computer climate models are consistent with widespread measurements across the globe, both on the land and at sea;[8] and (4) there are no other viable explanations because they have all been carefully and methodically ruled out. The last point—that all other explanations come up short—is nicely illustrated in a series of interactive graphics at a Bloomberg Businessweek online site, based on input from NASA scientists.[9]

In the meantime, atmospheric CO_2 has risen to concentrations higher than anything measured in bubbles of air trapped in ice cores that are up to 800,000 years old. The National Oceanic and Atmospheric Administration reports that "the last time the atmospheric CO_2 amounts were this high was more than 3 million years ago, when temperature was 2°–3°C (3.6°–5.4°F) higher than during the pre-industrial era, and sea level was 15–25 meters (50–80 feet) higher than today."[10] To put that date (3 million years ago) in perspective, *Homo sapiens* appear in the archeological record only about 300,000 years ago, so this is the first time since humans evolved that atmospheric CO_2 has been so high.

The reality of human-caused climate change is no longer debated at the mainstream scientific meetings of geophysicists, meteorologists, biogeochemists, atmospheric chemists, agronomists, foresters, and hydrologists. Rather, our discussions and research questions focus on why the warming is happening faster than we had expected even five years ago, how much the rate of warming might accelerate, and just how badly the economy and human health will be affected. It turns out that the majority of us were too conservative and too slow to acknowledge the rapid pace of human-caused climate change impacts, such as ice sheet losses in the Arctic and drought-enhanced forest fires that engulfed the western United States, Australia, and the Amazon forest in 2020 and 2021.[11] These climate change impacts are accelerating faster than we had understood even just 5–10 years ago.

Clear consensus statements issued by mainstream scientific societies demonstrate that despite the impressions to the contrary reported in partisan press and social media outlets, scientists with relevant expertise are now virtually unanimous about the gravity and causes of climate change. If you're not satisfied with the first opinion, for example from the National Academies of Sciences, Engineering, and Medicine,[12] then go for a second opinion from the American Geophysical Union,[13] a third from the American Meteorological Society,[14] a fourth from the World Meteorological Organization,[15] a fifth from the Agronomy Society of America,[16] or dozens more from similar mainstream scientific societies across the world. The World Meteorological Organization also coordinated a multi-organization, high-level compilation of the latest climate science available both as a traditional report[17] and as an online visualization.[18] For a summary of the impacts of climate change on human health, consider the joint statement of the editors-in-chief of the world's top medical journals:

> The risks to health of increases above 1.5° C are now well established. Indeed, no temperature rise is "safe." In the past 20 years, heat-related mortality among people over 65 years of age has increased by more than 50%. Higher temperatures have brought increased dehydration and renal function loss, dermatological malignancies, tropical infections, adverse mental health outcomes, pregnancy complications, allergies, and cardiovascular and pulmonary morbidity and mortality. Harms disproportionately affect the most vulnerable, including children, older populations, ethnic minorities, poorer communities, and those with underlying health problems.[19]

What Climate Change and Pandemics
Have in Common

Climate change and the COVID-19 pandemic are global phenomena that call for global responses. To be effective, those responses require scientific understanding of the causes, consequences, and cures. Fortunately for humanity, a scientific consensus emerged rapidly regarding the transmission of the coronavirus and intervention with effective and safe vaccines. Although the medical science journey was quicker to reach a consensus on COVID-19 transmission than was the climate change science journey, it is also a story of scientific integrity, as new evidence emerged and hypotheses were tested and retested. Unfortunately, the story and the pandemic are not over at the time of this writing, mainly due to resistance to the science-based remedies.

Initially, medical scientists were not sure which pathways of transmission of the coronavirus between people were most important. One hypothesis was that we mainly infect ourselves by touching surfaces contaminated with the virus and then touching our faces, thus transmitting the virus to our eyes, nose, and mouth. With surgical masks initially in short supply and not offering much protection from hand-to-eye contact, during the early stages of the pandemic experts recommended that surgical masks be reserved for health care providers. But as more data came in, and the original hypothesis was retested, it became clear that the dominant mode of transmission is through virus particles attached to droplets in our breath. Transmission by touching is still possible, so cleaning surfaces and hand washing are recommended, but the emphasis has shifted, based on the evidence, to preventing the spread of the virus through the breath of those who unwittingly are carriers and spreaders. Most cloth masks do not provide sufficient protection for the wearer from *inhaling* such particles, but they can catch most of the particles that the wearer *exhales*, thus protecting the people around the mask wearer in case that person is a virus spreader who does not yet have symptoms. Anthony Fauci, the director of the National Institute of Allergy and Infectious Diseases (and because of CO-VID-19 now probably the best-known living scientist in the United States and perhaps the world), changed his recommendation about facemasks as the scientific evidence accumulated. That is what good scientists do: when the evidence becomes overwhelming that a previously held hypothesis is wrong, a new working hypothesis is adopted. That hypothesis—that facemasks are effective in preventing the spread of the virus—has since held up well to further scientific scrutiny.[20]

Unfortunately, climate scientists and public health scientists now share another regrettable distinction. Both have become targets of the political far right. The right has ridiculed and dismissed climate science for several decades and for a variety of reasons discussed below. Now, medical professionals have also become targets of vicious personal attacks by those who, for various ideological, political, and selfish reasons, choose to deny the value of science and cast doubt on the gravity—and even the validity—of the coronavirus pandemic itself. Sadly, the reluctance to accept scientific evidence of the value of wearing facemasks and social distancing to limit COVID-19 transmission has evolved into a similar reluctance to accept the scientific consensus about the need, efficacy, and safety of vaccines. Misinformation about vaccines is not new, dating back to their invention, but time and time again, their value has been borne out. The simple fact is that those states and countries that have low vaccine rates for COVID-19, whether because of vaccine hesitancy in parts of the United States or lack of vaccine supply in developing countries, have been more susceptible to the wave of transmission, sickness, and death following the emergence of variants. While breakthrough cases occasionally affect the vaccinated (albeit usually with relatively mild symptoms), the pandemic that continues beyond 2021 is mostly a pandemic of the unvaccinated.

The Omicron variant and others yet to be discovered and named will evolve as the virus continues to spread among the unvaccinated, adding new challenges. Ultimately, the only way to get and keep this pandemic under control is through vaccines and the science that enabled them.

When Worldviews Lead to Denialism

A firm scientific consensus has made it possible to address both climate change and the pandemic through national government policies and international cooperation, provided that there is political will. The efforts also require individual, institutional, and private sector initiatives and commitments. In a word, they require *communitarian* solutions.

Communitarianism acknowledges that, while it is human nature for individuals to look out for themselves and their families first, communities of humans have always gained strength from working together. This was true for early human tribes of hunters taking down big game, for neighbors raising a barn, or for national armies defending sovereign territories. This cooperative instinct to protect one's kin and to help even distant relatives of the same tribe is due in part to an evolutionary process that improves the chances of passing down to the next generation the genes that family and tribal members share in common. Today, the threats that humans face go well beyond the adversarial animal or human that might be just beyond the next hill. Hence, our culture and politics now generally acknowledge the value of cooperating with broader and more distant communities for the sake of national security and public safety, and most recently for facing threats of global pandemics and climate change. Many communitarians also seek meaning for their own lives, both physical and spiritual, in helping others.

Few people quarrel with gestures of communitarianism so long as they are entirely voluntary, individual acts, or perhaps directed by their church, synagogue, mosque, or by trusted community leaders. But when individuals and communities, for whatever reasons, real or imagined, do not trust centralized government, then support often falls off for government policies designed to be in the public interest. In contrast to communitarianism, libertarians believe that it is only the individual who should decide what is in their best interest, not a government, even a democratically elected one, which they fear would trample their individual rights. This view is unworkable, in my opinion, especially for a world with about 8 billion people, soon to become 10-12 billion. The world is simply too full for that many unbridled self-interests refusing to cooperate in support of the common good. Government is also demonized, often disguised by the rhetoric of libertarianism, by wealthy and privileged individuals who wish to discourage government policies that might diminish their wealth and influence.

I am reminded of my friend Bruce's distrust of any sort of planning for how to manage what we now call the Anthropocene. Bruce's libertarianism drove him to reject not only most solutions that required government intervention, but also the very reality of problems that might require those kinds of solutions. Although climate change was only beginning to be discussed outside of scientific circles when I knew Bruce in the early 1980s in Zaire, he was highly skeptical of the science of the day that predicted both possible nuclear winter (years of cold weather caused by debris in the atmosphere after a nuclear war) and global warming (increasing temperatures caused by heat-trapping gases in the atmosphere). Looking for grounds to back up his suspicions that these were politically motivated, left-wing agendas, he repeated criticisms of the day that worries about both global warming and nuclear winter were contradictory. Scientists knew then and have accumulated further evidence since that there is no inconsistency between the phenomena of nuclear winter and global warming, because they are well-understood responses to two very different physical processes in the atmosphere. Massive explosions of thermonuclear weapons would eject huge quantities of ash and debris into the atmosphere, which would reflect much of the sun's energy back out into space, thus cooling the Earth. Although not as massive as a modern nuclear weapon explosion, volcanoes also inject large quantities of ash into the atmosphere, which clearly caused global cooling for a few years following the Mount Pinatubo eruption in the Philippines in 1992, so we know this effect is real. In contrast, greenhouse gases emitted into the atmosphere by burning fossil fuels act like an extra blanket, which traps more of the sun's energy within the Earth's atmosphere, thus warming the Earth. Nevertheless, as few as four years ago, I heard a high-level political appointee heading a US science agency express the same sort of doubt and denial as Bruce did decades earlier regarding the legitimacy of the climate change evidence produced by his own agency. Whether based on the libertarian principles or unabashed self-interest, these antigovernment views now promote science skepticism and denial in their public messaging. Why are libertarians and the self-interested wealthy only now widely denying science? As the human population has doubled in only the last 40 years, the world has become fuller and our challenges seemingly more intractable, making global cooperation more essential. Solutions to these challenges, like the pandemic and climate change, will require both scientific knowledge and communitarian cooperation. If those solutions threaten your worldview or your self-interests, then denying the science will likely accompany denying the communitarian solutions.

A personal experience regarding my own youthful worldview gives me a glimpse of how I might empathize with those who are skeptical and even those who deny the reality of human-caused climate change and the COVID-19 pandemic. When I was still in my thirties, my doctor noticed a faint heart murmur and told me to keep tabs on it. At that young, seemingly invincible age, I didn't worry. Not only was I young, but I instinctively assumed that serious health problems would befall other people, not me. Having a heart problem was not part of my worldview of personal possibilities. A decade later, while seeing my doctor about a bad cold, he noted that the murmur had grown louder, and so he ordered tests. I recovered from the cold, took the tests, and still didn't worry, because I was feeling great. After coming home from a rigorous cross-country skiing trip, I was in for the shock of a very inconvenient truth. The tests showed a leaky mitral valve, for which the only treatment was open-heart surgery. I was not only skeptical; I immediately went into denial. I had just finished a challenging skiing trip, I felt great, I wasn't yet 50, and I sure as heck wasn't ready to let someone cut open my chest and stop my heart to repair it.

My wife convinced me to get a second opinion, and we sought out another cardiologist with an excellent reputation. The results and advice were the same. I then did my own research, puzzling out the medical jargon of a review article in the New England Journal of Medicine; it was clear that people with my mitral valve diagnosis usually died within five years if they did not have corrective surgery. Eventually, my denial eroded away, overwhelmed by accumulated, undeniable evidence. I sucked it up and checked myself into the hospital. The heart surgeon stopped my heart, trimmed the extra flap that was preventing the valve from closing properly, revived my heart, and sewed me back up. I wouldn't wish that experience on anyone, but it was better than the alternative I was surely facing.

Sixteen years later I look back on that traumatic episode with four lessons learned: (1) denial is the easy way out, but its refuge is only temporary until reality sinks in; (2) bad things don't just happen to other people; (3) it can be very hard to do the right thing; and (4) expert medical opinion and medical care allowed me to overcome the denial and feel confident that I was taking the right course of action. Fortunately for me, there was no loud chorus of naysayers promoted by a 24/7 news and punditry cycle trying to convince me (falsely) that the cardiologists were still debating the cause or cure of my heart problem. The integrity and credibility of the well-respected, mainstream doctors and their profession were never in doubt. Had I kept searching for and found the one-in-a-thousand cardiologist who might have told me what I wanted to hear—that I didn't need open-heart surgery—I likely would not be here today, writing this book.

Similarly, if one looked hard enough, there might be a small fraction of medical experts who wouldn't recommend wearing facemasks or social distancing or getting vaccinated to protect against and prevent spread of the coronavirus, but heeding outlier advice rather than the scientific consensus has led to many hundreds of thousands of preventable deaths and debilitating long-term COVID symptoms. Similarly, deferring to the tiny fraction of scientists who may still question the human role in global climate change would lead to far more disaster, dislocation, suffering, and death. Regardless what message we may *want* to hear, we must listen to the messages that we *need* to hear, based on high-quality, mainstream science and medical expertise.

Like climate science, medical science has a history of building knowledge based on principles of chemistry and physics, hypothesis testing, and healthy skepticism, until multiple lines of evidence become compelling. My cardiologists used sophisticated optical imaging instruments and computers to estimate how much blood was leaking through my heart valve and used biology and physics to explain why I had sprung a leak. They called upon data gathered from dozens of studies to document my poor chances of survival if I did nothing, and they offered well-tested technologies and procedures to fix it. Likewise, the rapid development of a vaccine for COVID-19 was enabled by decades of biomedical research that tested and retested hypotheses about how the body fights viruses and how the virus's own RNA could be used against it in the design of an effective vaccine. Similarly,

Earth science has unequivocally measured how much the Earth is warming and has explained its human cause using observations and well-known and uncontroversial laws of physics. It has documented the biological and economic impacts of climate change and can point to the steps needed to minimize further turmoil and suffering.

I know from my own personal existential health crisis that denialism is by far the easiest response to anything that challenges one's worldview. When we don't want to give up our comforts and our familiar ways of looking at the world, then we simply deny those problems that our worldview is unable to solve. Whether it be in response to my youthful sense of invincibility, a libertarian's knee-jerk reaction to mask mandates for protecting public health, a greedy person's defense of their self-interests, or an isolationist's opposition to international cooperation to fight climate change, denial is an unfortunate reaction of the human psyche to be reckoned with. Denial seems to be the easy way out when confronted with evidence that human-induced climate change is undermining what makes our economy and our society tick, including agriculture, forestry, water supply, fisheries, and safe places to live and work. This is exactly what is happening with both climate change and pandemic denial, even in the face of millions of deaths caused by the pandemic and widespread fires and storms strengthened by climate change.[21]

This denialism then has a snowball effect, because to maintain it, the current strong scientific consensus and the credibility of top-notch, mainstream scientists must also be attacked. The scientists and those who embrace the science to formulate policy are branded as the other "tribe," while those who deny the science find comfort in the echo chambers of their favorite TV news channels and on social media. Effective media campaigns by climate change deniers and anti-vaxxers take advantage of and exacerbate this polarization. In addition to distributing misinformation, they often portray concerns about climate change as a recent fad of the political left, which includes intellectual, presumably liberal scientists from academia. The political left has its echo chambers, too, and it would be wise for them to listen in on other dialogs. However, characterizing climate scientists as an insular, left-wing group is simply inaccurate. In fact, I know climate scientists with widely ranging political views, from both major US political parties, from conservative to liberal, and also with a wide range of religious beliefs, from every major religion.

One of the most effective communicators about the science of climate change is Katharine Hayhoe, a highly respected climate scientist who happens to also be an evangelical Christian, born in Canada and living and working in Texas.[22] She listens to, empathizes with, and searches for common values with each of her audiences. She conveys the facts without talking down to people, and she shares her own personal experiences and values that others can respect, just as she respects theirs. Values can be discussed without making them politically charged, and the unequivocal *evidence* of human-caused climate change can be presented without political or religious prejudice. In her 2021 book, *Saving Us: A Climate Scientist's Case for Hope and Healing in a Divided World*,[23] Hayhoe recounts numerous stories of interactions with climate change skeptics by first searching for common ground and building trust on related topics, which then enables respectful and meaningful dialogs about climate change. Neither Dr. Hayhoe nor anyone can turn around some stubbornly held worldviews, but she gives us a model for communicating and defending the credibility of the science to those who will listen when they also feel listened to.

Denialism comes in different flavors and degrees. Polls show that some people think that the effects of climate change will occur far off in the future or that other people will be affected, not themselves (just as I thought as a young, invincible adult that heart problems only befall other people). That resistance is diminishing as the widespread effects of present-day droughts, heat waves, storms, and floods become more common, more extreme, and more palpable to more and more people. In the 1980s, there were only about three big climate-related disasters (huge forest fires, hurricanes, and other extreme weather events) per year in the United States that each cost over $1 billion, but in the 2010s, the average was about 12 billion-dollar-damage events per year. In 2020, there was a record setting 22 events with damages over $1 billion, totaling $95 billion that year alone. The cumulative cost of 285 such events since 1980 is about $1.9 trillion.[24] Some of these storms would have happened without climate change, but the trend of rapidly increasing frequency of really big climate-related disasters is one of the costs of climate change. Fortunately, the number of associated deaths has decreased, presumably due to improved weather prediction and learned behavior of people in those regions to evacuate when warned of impending risk. While people can get out of the way of extreme weather events, the properties, buildings, and infrastructure cannot, hence the staggering increase in damage costs.

Even without being a direct victim of climate-related catastrophes, the everyday impacts of persistent climate change on one's pocketbook can be especially persuasive. Consider the higher premiums for homeowner's insurance near coastlines, in floodplains, and in hurricane tracks;[25] reduced and inconsistent crop yields for farmers;[26] higher monthly water bills due to water shortages, and increasing public safety expenses and taxes for firefighting; or medical bills and loss of workdays due to heat stress.[27]

My heart condition did not go away while I was denying it. Ultimately, the facts caught up with me, and I had the choice of changing my worldview or risking premature death from a leaky mitral valve. COVID-19 and the pandemic were denied by politicians at the highest levels, and now we know their denial was responsible for a slow response and much unnecessary death, illness, and economic suffering. A coordinated, ongoing, and diligent effort will be required to slow and stop its spread before more variants evolve in this and in future pandemics. Nor will climate change simply go away. Extensive, persistent, and coordinated efforts are required at local, national, and international levels. As the grim realities and devastating impacts of climate change and pandemics catch up to us, will our society persevere in denialism and tribalism, or will we leverage our impressive scientific knowledge with neighborly goodwill, rallying to protect ourselves with communitarian alliances against common foes?

A COVID Consensus: No Shirt, No Shoes, No Mask, No Service

Across the political spectrum, "providing for the common defense" against invading armies and terrorists is a widely accepted and expected role for government. Providing for the common defense against invisible viral invaders, in contrast, does not seem to follow logically for many people. The libertarian or partisan view that government has no right to require individuals to wear facemasks in public places ignores the science that explains how the virus spreads. Worldviews aside, the fact remains that refusal to wear masks is a selfish choice that has made the pandemic much worse by increasing preventable transmission of the virus.

Most reasonable people agree that government has the responsibility to protect public safety by preventing the dumping of raw sewage, also laden with viruses and other vectors of human disease, into the rivers and lakes from which we draw drinking water. What is the difference between spewing out viruses in untreated sewage and spewing out viruses in unfiltered breath? Perhaps we have more respect for sewage because it usually smells worse than breath and because of its "yuck" factor, but good science during this pandemic has unequivocally demonstrated the odorless dangers lurking in each other's breath. Anti-mask partisans have wholly failed to explain why masks are an abridgement of rights, whereas sewage control is not.

Shops located near popular beaches often post signs requiring shirts and shoes to enter, which has not raised concerns of infringement of personal liberties. Why then, can the sign not say "no shirt, no shoes, no mask, no service"? Major retailers like Walmart, CVS, and Target attempted to require customers to wear masks in their stores during early phases of the COVID-19 pandemic, both for the protection of other customers and the employees, although enforcement was sometimes challenging. These private sector and government efforts have merely been trying to enforce the Golden Rule in public places during this pandemic: *breathe onto others as you would have them breathe onto you.*

Unfortunately, the overwhelming evidence regarding virus transmission and the safety of vaccines to fight it may not be enough, because even strong evidence without effective communication can fall short. We might take a lesson from an Italian virologist, Roberto Burioni, who gets right to the point. Appearing on a popular television show with opponents of vaccines, his retort to their anti-vaxxer views was: "The Earth is round, gasoline is flammable, and vaccines are safe and effective. All the rest are dangerous lies."[28]

I'm not sure whether Dr. Hayhoe and other climate change communication experts would endorse this tactic, but we may find a definitive statement like Dr. Burioni's useful for some discussions about the scientific consensus on human-induced climate change. Indeed, the science of climate change has progressed to the point where the original healthy skepticism of 50 to 60 years ago has been unequivocally answered. Although it may not be appropriate for relating to all audiences, we can now confidently apply Dr. Burioni's model for communicating the scientific consensus regarding the safety of vaccines to the scientific consensus about climate change: *Climate change is real; it is already happening; most of what has occurred in the last 50 years was caused by humans; it is already detrimentally impacting the economy, human health, and the environment; and there are affordable ways to combat it without hurting the economy. In addition, the Earth is round, gasoline is flammable, and vaccines are safe and effective. All the rest are dangerous lies!*

Just as leaders of many businesses have accepted the COVID-19 science and required mask wearing, many business leaders have also moved beyond climate change denial and are seeking genuine expert opinion on how to help slow climate change and to adapt to the climate change already underway. I watched CEOs of one big corporation after another stand up at the 2018 Climate Action Summit to pledge their company's contribution to reducing emissions and meeting the Paris Climate Accords target, even if the US government at the time refused to honor its own pledge. Insurance company actuaries have, of course, analyzed the increased risk exposures from rising sea levels and more severe storms. Property insurance premiums in low-lying coastal areas and other storm-prone areas have risen substantially, and some companies have stopped selling policies where risks are too high.[29] These aren't politically or ideologically driven decisions, but rather rational business decisions based on objective analysis of actuary data and risks.

Fortunately, these new perspectives by business leaders are supported by emerging evidence that there are economically and environmentally viable solutions to climate change, just as I had an option to fix my heart and just as vaccines have proved effective against COVID-19. These climate solutions are the result of yet another long journey in science and engineering to find alternatives to our reliance on fossil fuels, which is rapidly gaining momentum and offering hope, albeit with big challenges remaining.

Renewable Energy: A Journey from Pie-in-the-Sky Prospects to Mainstream Economics

As few as ten years ago, renewable energy was still contributing percentages in the single digits to the world's electricity supply. Wind and solar were still viewed by many as futuristic, pie-in-the-sky solutions to climate change. What a difference a decade makes. Utility-scale production of electricity from solar and wind is now more economical, without subsidies, than coal and oil. Markets are responding accordingly.

The central tenet of an effective strategy to slow and eventually stop climate change is to wean our economy off of its dependency on fossil fuels as quickly as is technologically, economically, and socially possible. Fossil fuels were a great ride, making many of us prosperous, and generally improving the quality of life since the beginning of the industrial revolution. But now, the bill has come due, as the harmful, unintended consequences of 160 or more years of emitting carbon dioxide and methane into the atmosphere by burning fossil fuels are coming back to haunt us. Therefore, it is time for a profound transition. Even the big oil companies understand that oil, gas, and coal will eventually be mostly phased out, although they would like to continue to ride the wave of a fossil fuel economy longer than humanity dare wait to avoid unacceptably harmful climate change.[30]

Of course, there will be economic and social disruptions linked to transitioning away from our carbon-based, fossil fuel–based economy, which will be discussed in the next chapter. However, the climatic disruptions and subsequent monetary and social disruptions caused by doing nothing would be far worse. Swiss Re, a reinsurance company that insures insurance companies, estimates that "the world stands to lose close to 10% of total economic value by mid-century if climate change stays on the currently-anticipated trajectory."[31] The pace of change from fossil fuels to renewable energy and the optimal mix of various forms of wind, solar, nuclear, hydropower, and dwindling fossil fuel contributions will be controversial. Nevertheless, the direction and the urgency of change are no longer in question. Similarly, the optimal policy approaches to achieve this deep decarbonization goal are debatable, but the status quo is clearly not acceptable, and change is already underway.

Economists tell us that the most efficient and effective way to decarbonize our society would be to put a price on carbon, such as a tax on consuming products that cause CO_2 emissions when they are produced or consumed. You and I would have more incentive to insulate our homes and drive fuel-efficient or electric cars if the cost of consuming energy generated from fossil fuel combustion goes up due to a tax on carbon. Ditto for owners of skyscrapers, trucking fleets, airlines, and everything else that uses fossil fuel energy.

The costs to consumers of a carbon tax would encourage everyone to make wiser energy choices. Over time, the carbon tax burden decreases for everyone as demands for low-carbon, low-cost technologies spur technological innovations that reduce their costs, enabling more of us to use these affordable alternatives and to use less fossil fuel. As we are weaned from our fossil fuel addictions, there will be fewer carbon products to be taxed. A good example of how this works is the declining revenues from cigarette taxes after those taxes were raised in the United States in 2009. There was a bump in tax revenue at first, but it has been declining since as more people were incentivized, in part by price, to quit smoking.[32] The revenues helped cover the cost of increased coverage under the State Children's Health Insurance Program, which, in combination with higher prices and complementary efforts by schools and nonprofit organizations, helped further reduce teenage smoking.[33] At the same time, regulations complemented the tax by nearly eliminating smoking in public places and thus exposure to second-hand smoke. Hence, neither taxes nor regulations nor educational outreach, alone, were enough, but together, these government and community actions had a synergistic positive effect on public health.

Increasing energy prices without concern for its impacts on consumers, however, may be socially and politically untenable, as the widespread recent, and sometimes violent, yellow-vest protests in France demonstrated. The French government's increases in fuel taxes coincided with discontent over the rising cost of living and increasing disparities of wealth, thus fueling the protests. In contrast, current proposals for carbon taxes in the United States are tied to using the funds raised from the tax to pay a dividend to households to offset their increased energy costs, so that the poor and middle classes are not unduly hurt.[34] The result could be revenue-neutral for the government levying the tax, while payments back to those consumers hardest hit by the increased prices would ease their pain.

Indeed, a unique feature of the carbon tax is that it provides a mechanism to raise money to help the most vulnerable people, who will suffer the most from climate change. The wealthy and the upper middle class have more means to adapt to climate change by moving away from areas prone to flooding and paying more for food, air conditioning, and water consumption. As we discussed in chapter 2, disparities of wealth are growing in nearly every country of the world, as the upper economic classes have leveraged their positions and prospered more than the lower economic classes. Addressing the causes of growing wealth disparity will require more than a carbon tax, to be sure, but dividends from the carbon tax revenues could significantly help the people who are most vulnerable to the impacts of climate change.

On the flip side, some opponents to a carbon tax from the left, including many green new deal supporters, argue that the wealthy should not simply be allowed to pay more to pollute. They prefer regulations, applied equally to everyone, that prevent pollution and punish polluters. They also believe that reductions in greenhouse gas emissions resulting from a carbon tax would be too incremental, because it does not cap emissions directly and immediately. Opponents on the right, in contrast, predictably recoil from the T-word, arguing that any additional taxes will foster more big government, even if the tax is structured to be revenue-neutral by rebating most of the revenue back as dividends to those who would be hit hardest by the tax (like the yellow-vest protestors in France). They also contend, without supporting evidence, that a tax on forms of energy and transportation that produce CO_2 emissions would stifle the economy. To the contrary, evidence from those places where carbon taxes have already been employed demonstrates that subsequent stimulation of ingenuity and technology to avoid emissions actually promotes those sectors, and the feared economic setbacks do not materialize.[35]

Experts agree that a carbon tax alone would not be a panacea, because market-based responses, while often powerful, are not always enough, and other strategies can be more effective in some cases. Regulations that mandated improvements in the average fuel economy of a manufacturer's auto fleet, for example, have proved effective to incentivize auto companies to develop and market more fuel-efficient cars and trucks. At the same time, while the average fuel economy improves, some models will always be more fuel-efficient than others, so why not also apply a carbon tax to gasoline to make the owners of the less efficient vehicles pay more to operate them? This is a form of punishment for them, an incentive for others to buy the more efficient models, and a means to raise revenues to assist those who may have trouble purchasing the more expensive fuel. Personally, I think we need an all-of-the-above strategy, each applied where it makes the most sense and is most effective.

The transition from fossil fuels to renewables will result in a net increase in jobs nationally and globally,[36] but the story will be different in some specific localities. Help will be needed for those families in regions that lose jobs due to constriction of the coal, gas, and oil industries. Such disruption will happen in any case, as automation in mining and other sectors continues to replace humans with robots, regardless of energy policy. We should be honest, however, that the decarbonization that is so urgently needed to avoid catastrophic climate change will accelerate the decline of the coal industry (already long in decline in the United States for other reasons discussed in the next chapter), with the oil and gas industries soon to follow. Facing that dilemma will require convergent solutions involving business, union, and community leaders, social scientists, economists, and politicians. The tensions caused by changing technologies and demands of society, including the ongoing decline of the coal industry, is the topic of the next chapter.

In lieu of a carbon tax, a cap-and-trade system is another option that would put a price on carbon. Cap-and-trade was used very effectively in the 1990s to reduce sulfur emissions that cause acid rain.[37] The government set nationwide caps on total sulfur emissions from industry smokestacks and issued emission permits to the companies responsible for most of the emissions. Those companies could trade those permits on an emissions trading market. The innovative companies that had reduced their emissions below their permitted level were rewarded with an extra profit when they sold their unused emissions permits, while the less innovative companies that needed to buy more emission permits were penalized by the additional cost. Over time, the government reduced the number of allowable emissions, so that total emissions came down, but the government never dictated which companies had to make the reductions or what technologies they had to use. Instead, the marketplace sorted out the winners and losers and provided incentives to those who devised innovative solutions to lower their emissions. Critics of acid rain regulation had warned that it would be too expensive, but the cap-and-trade approach ended up saving innovative companies a lot of money while very effectively reducing sulfur emissions.

Using the same approach to reduce CO_2 emissions will be more challenging, because there are many more emitters from many sectors of the economy. Yet this is doable, with many lessons already learned from successes and failures in the several countries and states that have tried cap-and-trade markets.[38]

Our society has become accustomed to regulations in some sectors, such as local and state building codes and fire codes to help ensure that occupants are safe. Similarly, new building codes should ensure that buildings are designed to be highly energy efficient and that old buildings are retrofitted with energy efficiency technologies. My son lived in a recently renovated walk-up apartment in New York City, but the owners failed to retrofit the heating system during the renovation. As a result, he ran his window air conditioning unit in the winter because he had no way of turning down the excessive heat in his apartment. I cringe to think of the energy wasted. The renovation must have been permitted by the city, which was a lost opportunity to improve energy efficiency. Fortunately, New York and other cities have passed aggressive legislation to cut greenhouse gas emissions from large buildings by about half by 2030 and by 80% by 2050.[39] Emissions from the iconic Empire State Building have already been reduced by 40%, and the target is for it to be a net zero-emission building by its 100-year anniversary in 2030.[40] This sets a high bar, literally and figuratively, for historic buildings of previous centuries to be brought up to the needs of the twenty-first century.

As already noted, mandated fuel economy standards for automobile manufacturers have been very effective at stimulating innovation, reducing emissions, and saving money at the gas pump for consumers. Similarly, regulations on gas and oil companies were starting to successfully reduce pipeline leaks of methane, a potent greenhouse gas contributing to climate change, much of which comes from the leaky oil and gas supply chain.[41] Those regulations were repealed by the Trump administration and then reinstated by the Biden administration. The biggest natural gas companies have the capital to invest voluntarily in technologies that reduce methane, and they do so because plugging leaks of valuable natural gas pays off for them in the long run. The smaller companies, however, need either cost-sharing incentives or regulations or both to check their leaks of that potent greenhouse gas. The United States was one of more than a hundred countries that pledged at the November 2021 meeting in Glasgow, Scotland, of the Twenty-Sixth Conference of the Parties (COP26) of the United Nations Framework Convention on Climate Change to reduce methane emissions by 30% by 2030.[42] Now we must see if those countries will live up to their pledges and whether the holdouts with large methane emissions—China, India, Russia, and Australia—can be convinced to come along.

Incentives to reduce greenhouse gas emissions can also take the forms of reduced taxes, low-interest loans, or rebates to encourage homeowners and building owners to install energy-saving technologies, such as green roofs, reflective roofs, solar panels, and batteries for storing energy. Green roofs with verdant gardens not only reduce a building's energy use by dissipating incoming heat from the sun; they also reduce storm runoff, offer an urban refuge for bees and other insects and birds, and even provide space to grow tasty tomatoes (and perhaps my favorite yummy Brussels sprouts). Perhaps we should not oversell the potential to grow food on rooftops in cities, but all co-benefits merit consideration in addition to the substantial potential for reflective or green roofs to save energy. As demand for these products grows, engineers and entrepreneurs will harness their ingenuity to offer better products at lower prices, as we have already seen for solar panels and wind turbines in the last decade.

I took advantage of tax rebates to offset some of the initial costs of installing solar panels on the roof of my house and for buying an electric car. Most of my local transportation is not only gasoline-free, but also driven by photons captured by the solar panels on my own roof! Payback on my initial investment will take several years. Not everyone can afford the initial investment now, but the installation costs of residential solar have dropped by more than half since I installed solar panels in 2011 and bought the electric car in 2013, so more and more people will be able to opt in to these remarkable technologies.[43] The tax incentives and rebates are important for early adopters, and our purchases helped stimulate the market, which led to lower prices, thus creating a snowball effect of broader and broader participation in these low-carbon technologies. As shown in figure 5.1 the cost of utility-scale solar and wind energy has dropped even more, to the point where the marketplace, without subsidies, is now driving corporate decisions to invest in new renewable sources of electric generation over fossil fuel sources. Although this is

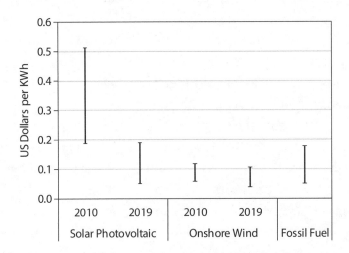

FIGURE 5.1 Cost of Commissioned Utility-Scale Power. The bars represent the 5 and 95 percentiles of global levelized cost per kilowatt hour (KWh) of energy for newly commissioned utility-scale solar photovoltaic and onshore wind production in 2010 and 2019 (in 2019 US dollars). The range of fossil fuel-based electricity production is shown for comparison. Drawn by the author from estimates by the International Renewable Energy Administration, *Renewable Power Generation Costs in 2019* (Abu Dhabi: IRENA, 2020), https://www.irena.org /publications/2020/Jun/Renewable-Power-Costs-in-2019

movement in the right direction, it is not happening fast enough. These market forces still need a push.

Electrification of transportation is another basic tenet of a green new deal, which will require policies that incentivize consumers, manufacturers, and engineers to develop and adopt consumer-attractive electric vehicles (EVs) and abundant charging stations. After that initial boost, EVs will become mainstream and will advance by market forces, just as renewable energy is now economically favorable.

While these trends are very promising, the challenge of decarbonizing our society as quickly as possible is immense. It is beyond the scope of this book to propose the specific combination of market-based, incentive-based, or regulatory approaches needed for each sector and subsector of the economy. The Biden administration is currently not pursuing a carbon tax or a cap-and-trade option per se as part of its climate change agenda. Instead, its initial proposal was a combination of approaches: it would have incentivized electric utility companies to increase electricity generation from renewable sources with a set of rewards for reaching targets and fines for missing them; it avoided strict regulatory mandates, but the penalties for not meeting targets could be significant; and it gave utilities some flexibility to take advantage of market opportunities to figure out the most efficient and cost-effective way to achieve renewable energy targets. The rewards for reaching targets would also be shared with consumers, thus helping keep costs down for low-income households. The plan also proposed incentives for auto manufacturers and consumers to transition rapidly to electric vehicles. At the time of this writing, a compromise between the president and key members of Congress had not yet been reached.

We will need some judicious combination of these proposed carrots and sticks and market-driven technological developments, both now and in the future. The choices will not be easy and the debates will be vigorous because they will explore both what is technologically possible and what various stakeholders value. I believe that a price on carbon, either through a tax or through cap-and-trade or through incentives and penalties, will ultimately be necessary to achieve the needed deep decarbonization, but I support other solutions as well. In addition to national policies, we need a policy environment in which local innovation by communities, businesses, and individuals can flourish with broad engagement of many stakeholders.

Congressional action in 2022 will by no means be the last word; the United States and other countries will require additional steps to meet their Glasgow COP26 pledges and more. Those pledges were not enough, as recognized by the final agreement that the countries of the world must return in 2022 and subsequent years with more ambitious pledges by the wealthy nations to cut emissions further and more quickly, to meet past pledges of providing $100 billion annually to finance mitigation in poorer countries, and to at least double funding to those countries to help them adapt to climate change already underway. The private sector also stepped up at the COP26 meeting to pledge more than $130 trillion to transform the economy to net zero emissions, which is estimated to be about 70% of what is needed to get the job done, again assuming that pledges are honored.[44] As with the other pledges at COP26 to reduce methane emissions by 30% and stop deforestation by 2030, the thousands of young people demonstrating at the conference in Glasgow expressed skepticism, bordering on disbelief, that their elders would truly honor these pledges. "Prove us wrong," pleaded Vanessa Nakate, a young climate activist from Uganda.

The Four Pillars of Deep Decarbonization

A number of remarkable reports on "deep decarbonization" were published within a few months of each other in early 2021, including from the International Energy Agency,[45] the US National Academies of Sciences, Engineering, and Medicine,[46] and a peer-reviewed paper in the scientific journal *AGU Advances*,[47] for which I had the honor of acting as the editor. All three give very similar take-home messages. The latter analyzed the feasibility of the "deep decarbonization" of the US economy needed to reach zero net CO_2 emissions by 2050 and concluded that it can be done at a net cost of 0.2%-1.2% of GDP. Increased spending on capital costs for low-carbon, efficient, and electrified technologies is estimated at almost $1 trillion in 2050, but it is offset by nearly as much in savings from reduced spending on fossil fuels and related technologies. The net cost is estimated to be about $1-$2 per person per day, which is far lower than previous estimates. This reduction in cost is largely due to decreased costs of solar and wind power (see fig. 5.1) and batteries. The estimated savings are an underestimate, because they do not include the savings from the costs to society of the avoided climate change described in the report by the Swiss Re Institute noted above (for example, avoided damages and deaths from storms, droughts, heat waves, floods, crop losses).

The successful pathway to deep decarbonization of our economy, which meets energy needs at reasonable costs, depends upon four pillars. The first includes the many examples of energy efficiency discussed earlier in this chapter. We must continue to find ways to use energy more efficiently for lighting, heating and cooling, transportation, appliances, and manufacturing. The next three pillars also may not be entirely new to most readers, but this report has some new twists.

A second pillar requires that we convert most of the current end uses for fossil fuels to technologies powered by electricity. With a few exceptions discussed below, electricity is more thermodynamically efficient for most applications, and, most importantly, it can be generated through renewable sources of energy. This means heating and cooling homes and buildings with high-efficiency heat pumps, electrifying most automobiles, and converting many industrial processes to electricity where it is thermodynamically favorable. These technologies already exist, but they need incentives to boost their adoption more broadly. The governments of the world currently provide subsidies to fossil fuel use worth over $400 billion,[48] which should be eliminated or shifted to promoting uses of electricity generated by renewables. By 2030, we need electric vehicles and heat pumps to make up over 50% of new purchases. A side benefit of converting vehicles, heating systems, and industries to electricity is that smog-producing emissions will drop alongside the decline in CO_2 emissions, resulting in cleaner air and less respiratory disease and heart disease. The cleaner air that we observed during the economic shutdown in early phases of the pandemic in 2020 could become permanent if we adopt these cleaner technologies. Thus, a silver lining of the pandemic's temporary economic shutdown is that it showed us what is possible for air quality, and that we needn't take for granted that air pollution is inevitable.

The third pillar follows from the second, because if we are going to convert most things to electricity, the electricity had better be generated from mostly carbon-free sources, primarily from renewable resources. The report calls for 80%–90% of electricity to be generated by solar, wind, hydropower, and geothermal sources by 2050. This conversion to renewables is already underway with existing technology (remember, it contributed less than 10% only about a decade ago), but more incentives are needed to accelerate the pace to get to 80%–90% by 2050. In the next decade, we must more than triple renewable energy capacity to get on track for that 2050 goal.

Interestingly, there will continue to be a role, albeit declining over time, for natural gas and existing nuclear power to fill gaps in energy production by renewables and storage. The report concludes that an electricity system using 100% renewable energy would be technologically feasible in 2050, but the transition of that last 10% or so would be very expensive. Because wind and solar production of electricity is not continuous, and because storage of energy in batteries is also currently limited to a few hours of demand, there will continue to be a small role for natural gas and nuclear power generation of electricity through 2050. However, those plants would operate less than 15% of the time— just enough to provide round-the-clock reliability of the electricity supply when the sun is not shining, the wind is not blowing, and the battery storage has been used up.

It is possible that a new generation of small-scale, modular, advanced nuclear power plants might also be developed to contribute to the round-the-clock reliability. However, as described in the National Academies report cited above, that new nuclear technology would have to be more cost-competitive than the past and current generations of large-scale nuclear reactors, which are far too expensive to justify building anew. The safety of nuclear waste disposal has been a persistent technological and political challenge for nuclear power, which may ultimately be its Achilles' heel, even for new generations of reactors. The decarbonization scenario does not depend upon nuclear in the long term, but R&D on a new generation of smaller-scale, modular, advanced nuclear reactors and on the technical and societal concerns about waste disposal should continue.[49] The jury is still out, but this new generation of nuclear power technology might contribute to a future mix of reliable, affordable, and safe energy sources.

In certain industries, such as steel, cement production, aviation, and maritime shipping, developing technologies for substituting fossil fuels are still in a stage of emergence. These will also likely require use of fossil fuels for a few more decades.[50] These are among the few exceptions to short-term electrification mentioned in pillar 2. Cement production accounts for about 4% of global CO_2 emissions.[51] Concrete and cement products that produce somewhat less CO_2 are already being marketed, albeit still at premium prices and not with as much carbon reduction as needed.[52] Aviation fuel derived from recycling vegetable oils and from plant sources is used in relatively small quantities now, and also at premium prices. These industries are yet another area where R&D to reduce emissions and ultimately to decarbonize is urgently needed.

Fourth, more R&D is also needed on technologies called "carbon capture and storage." These technologies absorb CO_2 from smokestacks, from burning biofuels, or from making cement, and then safely bury that captured CO_2 deep in the Earth. These technologies also already exist, but they are currently not economically competitive in most situations, so the R&D must focus on bringing these costs down. Carbon capture and storage is not a substitute for the other three pillars, but it will allow some residual use of fossil fuels in the cement, steel, aviation, and maritime shipping sectors while new technologies are developed. It will also allow some natural gas combustion to fill in gaps where renewable or stored energy does not meet around-the-clock demands. Beyond 2050, carbon capture and storage may also play a beneficial role to eventually draw down atmospheric CO_2 concentrations back toward the levels that existed prior to the industrial revolution, so that we can reverse some of the damaging climate change that has already occurred.

Interestingly, some big oil and gas companies have been investing in R&D on carbon capture and storage, which suggests that they are hedging their bets for when society requires that it be applied to any future use of fossil fuels.[53] We should not be naïve—their emphasis on carbon capture and storage may well be a delaying tactic to sidetrack rapid decarbonization policies that would reduce their oil revenues—but their progress on this needed technology will become useful even with rapid decarbonization. Those companies have experience and know-how, so they could become part of the solution.

The last hours of the COP26 negotiations in Glasgow focused on whether the final agreement would include a statement that coal power and government subsidies for oil and gas should be "phased out." Astonishingly, this was the first COP agreement that explicitly mentioned fossil fuels. India argued that it does not have the capacity to phase out fossil fuels completely unless the wealthy nations step up with very significant financial aid to enable that transition. Hence, "phase down" fossil fuels was the final wording compromise, which disappointed many activists and especially the governments of island nations, like Maldives, that will soon be inundated by rising seas. I understand their frustration, as the wording is highly symbolic of the degree of ambition. The more important question in my mind, however, is how quickly we can reduce the use of fossil fuels to a small fraction of current levels. When the last 10%–20% of current fossil fuel usage can be phased out completely is less important than how quickly the first 50% can be phased out and then how quickly the next 25% can be eliminated, and so on. On that score, too, the Glasgow Agreement was insufficiently ambitious,[54] but it established a transparent reporting process to hold countries accountable to their pledges. All citizens, young and old, must keep up pressure on governments and the private sector for meeting and expanding their pledges.

Forests Are a Big Part of the Solution Too

The deep decarbonization study published in *AGU Advances* did not directly address forest and soil management as carbon sinks, but as discussed in chapter 4, regrowing forests on cleared land that was only marginally productive for agriculture is another worldwide strategy for taking carbon out of the atmosphere. However, this approach, called "Nature-based Climate Solutions,"[55] is not as simple as saying that we will plant a trillion trees, as some politicians have suggested. Competition for land with agriculture limits how much land can be reforested. A best-case scenario for natural forest regrowth on degraded lands could draw out of the atmosphere about 20% of current annual global CO_2 emissions from fossil fuel combustion.[56] This assumes that we first stop the current practice of tropical deforestation, which is responsible for about 15% of current annual global CO_2 emissions.[57]

Brazil had made great progress on this front, at least temporarily, by reducing its rate of Amazon deforestation by a factor of more than five. At its peak in 2004, almost 28,000 square kilometers were deforested (about 10,000 square miles; about the size of the state of Maryland deforested in a single year). By 2012, Brazil had reduced its annual Amazon deforestation to less than 5,000 square kilometers (1,900 square miles).[58] This impressive decline in deforestation rates was possible because of a combination of policies, including incentives from supply chains that required sourcing agricultural commodities from areas not recently deforested, implementation of improved deforestation detection from satellite images, governmental commitment to enforcement of laws, protection of indigenous lands, and innovative finances for rewarding forest cover. At the same time, Brazil's agricultural exports grew, demonstrating that its agriculture and economy were not dependent upon cutting down more and more forest.[59] Many farmers and ranchers were committed to increasing their agricultural productivity while also protecting the remaining forests on their properties. Since then, however, changes in the environmental laws, lax enforcement, and an effort to open up indigenous lands to agriculture by the administration of President Jair Bolsonaro have resulted in the deforestation rate climbing back up to nearly 10,000 square kilometers in 2019.[60] Brazil had been on the path of demonstrating to the world that economic development and forest conservation could go hand in hand. Convergent knowledge of forest, agriculture, and conservation management could make it a world leader once again if there were the political will to do so. Brazil joined over 130 other countries at the COP26 meeting in Glasgow to "commit to working collectively to halt and reverse forest loss and land degradation by 2030 while delivering sustainable development and promoting an inclusive rural transformation."[61] Those countries collectively cover more than 90% of the world's forests, so this is a very good sign, but of course, the commitments will have to be backed up with actions.

Forest management and conservation are part of the climate change solution, and will also yield benefits of biodiversity conservation and other ecosystem services, but we shouldn't be fooled by attractive-sounding policies of simply planting several billion or even a trillion trees. That sounds great, but where will they be planted, and will they be cared for so that they survive and grow well? Will they compete with land needed for growing crops or for groundwater needed for agriculture, human consumption, and fisheries? Will they be diverse mixtures of tree species that support a diverse community of other plants and animals, or will they be monoculture plantations? Are local communities on board to help nurture the trees and conserve the forests? Planting trees can be admirable, but managing forests for carbon sequestration and other benefits involves much more than just planting trees.[62] I welcome such proposals as long as there is also a well-developed and comprehensive plan for forest management and the necessary community engagement.

Even Renewables Have Drawbacks

Perhaps you have heard that wind turbines make noise and kill birds. Fortunately, a convergence of engineering, social science, epidemiology, bird and bat ecology, and engagement with communities has emerged to find acceptable solutions. The technology of wind turbines has come a long way from the old-fashioned windmill used to pump well water before electricity was available in rural areas. The technology could be honed further, making the blades more aerodynamic and constructed of materials that have less impact on the environment. Technology could also improve the way the intermittent wind-generated electricity is integrated into the electrical grid and stored. But these technological advances alone, while important, will not address one of the biggest impediments to further spread of this form of renewable energy, which is pushback from communities that are skeptical about or even strongly opposed to their presence.

Proponents of renewable energy technologies should be no more immune to questions about unintended socio-economic and environmental impacts than are the coal and oil companies about their products. A green new deal cannot simply insist on a technology because it is renewable and "green" by someone's definition. The economic and social impacts of green technology are just as important as the environmental justice impacts of air and water pollution generated by fossil fuels. Some of the concerns raised by local communities about wind turbines have more merit than others, but all must be addressed.

As I was house hunting prior to my move to western Maryland, we found a house that we really liked in a wooded area not far from one of the many Appalachian ridgetops. One of my future colleagues who lived in that same neighborhood warned me that a line of several wind turbines was about to be installed on that ridge. When I returned a few months later on another house-hunting visit, the giant turbines were already there. I didn't mind the visual effect, but sitting in my parked car with the window rolled down in front of the house for sale, I could hear the constant swish-swish of the rotating blades. Imagining what it would be like trying to fall asleep on a summer night with the windows open, I didn't bother to get out of the car and take a second look at that house. I suspect that those homeowners had trouble selling their house and probably had to take a lower price because of the nuisance that the noisy wind turbines introduced.

There have also been claims that wind turbines harm human health, but they are not supported by epidemiological studies. On the other hand, the noise that I found merely annoying could be bad enough in some extreme situations to affect mental health. The noise issue alone is enough for me to side with local residents that some sort of minimal distance should be required, or the turbine company should have to buy out or fairly compensate the homeowners for the damages caused by the impacts of the turbine location. Setting the distance and noise thresholds will be controversial and challenging, but good convergent science, transparent communication, and advance planning can reveal the technical, human health, and social science evidence while also engaging the local stakeholders.

Many people are also understandably concerned about the impacts of wind turbines on birds and bats. This is a topic on which my colleagues at the Appalachian Laboratory of the University of Maryland Center for Environmental Science have considerable expertise to measure the sizes and distributions of bat and bird species killed by wind turbines.[63] It turns out that we know a lot about which species of bat are at risk and when.[64] Only the migratory species of bats are killed in large numbers; they migrate only during a few weeks each year; and they fly only when it is not too windy. Turbine operators can and have agreed to turn off the turbines at certain wind speeds during nights of the bat migratory season, which greatly reduces bat mortality. Some revenue is lost from this modest curtailment of energy production to avoid bat mortality, but that is a reasonable trade-off. Fortunately, good convergent science identified an effective and reasonable compromise.

Similarly, migrations are important for many bird species, and understanding when and where those migrations occur can lead to better management of the turbines to reduce bird mortality. The risks to birds from turbines, however, must also be placed into a broader context. By far the biggest killers of birds are cats, both domestic pets that are allowed outdoors and feral cats. These cats not only kill billions of birds in the United States alone; they also kill mice, lizards, and newts, some of which are rare and endangered species. I am a cat lover myself, but I learned to keep my cat indoors, both for her safety and that of the animals around my house. My screened-in patio was also her "catio," allowing her to be entertained by sights and sounds of the outdoors, but keeping her safe from the coyotes, and the mice, birds, and amphibians safe from her. More urgent than mitigating the much smaller number of birds killed by wind turbines (for every bird killed by a wind turbine, ten thousand birds are killed by cats; see fig. 5.2), we need a big awareness campaign to urge owners to keep kitty indoors, as well as devote resources to campaigns to neuter or euthanize feral cats. Anecdotally, when I recently hiked in a

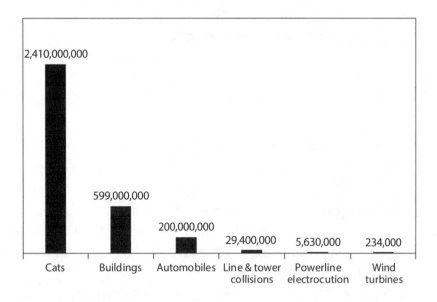

FIGURE 5.2 Causes of Annual Accidental Bird Mortality in the United States and Canada. Redrawn by the author from data in Scott R. Loss et al., "Direct Mortality of Birds from Accidental Anthropogenic Causes," *Annual Review of Ecology Evolution and Systematics* 46 (2015): 99–120, doi: 10.1146/annurev-ecolsys -112414-054133

mountain range in China, I was surprised by how few birds I saw and heard, and by how many feral cats I encountered. Being only a tourist, I cannot confirm the connection, but I strongly suspect that this problem is common throughout the world.

Climate change is another important threat to birds. Some bird species sense when it is time to migrate based on the length of the day, which was no doubt a strong evolutionary force for their survival. However, their food sources, whether they be plants or insects, also respond to seasonal changes in temperature. Temperature is changing with climate change, but the seasonal pattern of day length remains unchanged. Hence, some birds that time their spring migration by day-length cues may arrive at their summer homes too late to find their required food sources if global warming has caused the needed phases of plant and insect growth to have already come and gone. By phasing out fossil fuels and thus mitigating climate change, the expansion of wind turbine technology would be a net benefit for most bird species, despite the loss of some birds in some locations.

Batteries Are the New Oil

While wind and solar get most attention these days, one of the key technologies that will be needed for deep decarbonization pillars 1–3 is the development, expansion, and smart deployment of a new generation of batteries that store energy from renewables and that enable electrification of automobiles capable of driving long distances on a single charge. In some developing countries, distributed renewable energy with sufficient battery storage could leap-frog the old-fashioned reliance on large, centralized electricity generating facilities.

Battery technology is developing rapidly, and costs are plummeting, but further development of affordable, effective batteries is one of the key technological challenges of this transition. For short-haul delivery trucks, taxis, and other cars that put on many miles per year, the total cost of owning an electric vehicle is already considerably lower than a gasoline vehicle. Amazon, UPS, and FedEx, for example, are moving aggressively to electrify their delivery truck fleets.[65] Long-haul trucking will be slower to electrify and will need development of more efficient batteries or fuel cells that produce hydrogen. Daimler plans to sell battery-powered short-haul trucks that can compete on cost with diesel by 2025 and has a goal of selling long-haul fuel-cell trucks that are competitive with diesel by 2027.[66]

A major transition is also underway for cars. The batteries in the current fleet of electric cars cost about $180 to $200 per kilowatt hour. For electric cars to be cost-competitive for the average car owner who drives about 12,000 to 15,000 miles per year, the cost of batteries needs to be cut about in half, to $100 per kilowatt hour.[67] They must also support a 300-mile driving range without charging, and charging must be relatively quick. A Bloomberg analysis predicts that, even without subsidies, electric vehicles will be at price parity with most types of gasoline and diesel vehicles by the mid-2020s.[68] One catch, however, is that most people underestimate by about 50% the costs of operating a gas-powered vehicle. Hence, they tend to be more affected by the initial purchase price than the lifetime operating costs when making a decision on what model to buy.[69] Electric vehicles cost more up-front to buy but much less to operate annually. They need less maintenance (no major oil changes or spark plug tune-ups), and electricity usually costs less than gasoline per mile driven. This confusion between initial versus lifetime costs could be addressed with sales tag labeling that projects the lifetime cost of owning and operating the vehicle, such as the labeling currently used in the United States on washing machines, refrigerators, and other major appliances. Rebates and tax deductions would also help consumers swallow the higher up-front purchasing costs.

In the meantime, China sold over 1 million electric cars in 2018. Indeed, China is well positioned to dominate the global markets of electric cars and batteries, but American and European manufacturers could still catch up. General Motors, Volvo, and Jaguar have pledged to sell only electric cars by 2035 or sooner. Ford is investing heavily in EVs, introducing its all-electric F-150 pickup in 2022. In addition to China, several European countries, California, New York, and Washington state intend to ban the sale of gasoline-driven and diesel-driven cars by sometime between 2025 and 2040. A combination of policy initiatives, technology development, private sector commitments, and favorable market forces indicates that a major transition to electric vehicles is in store very soon and certainly within the next two decades.[70]

As this transition continues, imagine its impact on geopolitics, as oil-producing nations in the Middle East and Russia and multinational oil companies lose markets and, hence, political clout. The geopolitics of oil may be replaced by a new geopolitics of lithium and rare earth metals. Lithium is a key component of what appears to be the most promising battery technologies for the future. It may become the new limiting resource that affects geopolitics, because relatively few countries possess most of the world's reserves of lithium.[71] Unlike oil, however, the lithium in batteries is not consumed, so it can be recycled. Indeed, recycling technology will be essential for the sustainable viability of future battery technology. This is but one example of the needed circular economies that we will consider in chapter 7 (along with the significant environmental hazards associated with mining operations for lithium and rare earth minerals).

These are ambitious, but entirely realistic goals, technically, economically, and socially according to the deep decarbonization study:

> By methodically increasing energy efficiency, switching to electric technologies, utilizing clean electricity (especially wind and solar

power), and deploying a small amount of capture carbon technology, the U.S. can reach zero emissions without requiring changes to behavior.[72]

I'm not so sure that some behavioral changes won't be needed, or at least helpful to meet deep decarbonization goals, but the point is that neither the monetary nor behavioral costs need be unacceptably burdensome. In fact, the 0.2%–1.2% GDP cost of deep decarbonization of the energy sector will also yield welcome benefits, such as less respiratory illness and heart disease due to cleaner air and less costly climate change adaptation measures (for example, fewer seawalls to protect against sea level rise and storm surges, less frequent major flood damage, fewer heat waves and droughts, fewer losses due to wildfires). We are already committed to significant climate change that will require some expensive adaptation costs, but investing in deep decarbonization now will reduce the amount of expensive adaptation in the future.

Powering with Willpower

When I was coming to grips with my heart problem, I went to the best cardiologists and surgeons that I could find. Patient Earth and her inhabitants also have excellent and dedicated scientific experts, policy wonks, and engineers to diagnose and treat the Earth's climatic fever. Taking expert advice is not always easy, I know from personal experience, but the alternative is much worse. We already have many of the technologies and knowhow needed to wean ourselves from fossil fuel addiction, and the fruits of further innovation in private and public sectors, driven by a variety of incentives, await discovery.

Knowing that the needed expertise and technology is at hand and affordable in most cases is encouraging, *but* technological advances alone will not be enough. The willpower of consumers, voters, community leaders, business leaders, and policy makers to stop denying and to start taking actions will also be essential. Listening to the young people marching for climate change action, it is becoming clear that a growing number of the Earth's human residents, many with wisdom beyond their years, understand that we are not invincible to climate change. Indeed, they have the desire and willpower to seek the dramatic changes, like those envisioned by a green new deal. Perhaps it takes a schoolgirl on the autism spectrum, like Greta Thunberg, and many less famous kids all over the world to make the adults acknowledge that we have been denying the reality of their generation's present and future well-being. These young people demand that we suck it up and act on the message that we know is right rather than deny and procrastinate based on disinformation peddled by those who seek to perpetuate their own worldviews, self-interests, and power.

Recommendations

The Green New Deal (H. Res. 109, 116th Congress, 1st Session [introduced February 7, 2019]) establishes three clear goals to address climate change mitigation and impacts:

> to achieve net-zero greenhouse gas emissions through a fair and just transition for all communities and workers; . . .
>
> to invest in the infrastructure and industry of the United States to sustainably meet the challenges of the 21st century;
>
> to secure for all people of the United States for generations to come—
> (i) clean air and water; (ii) climate and community resiliency;
> (iii) healthy food; (iv) access to nature; and (v) a sustainable environment.

It then offers a list of initiatives (A–N below) proposed for the next ten years to meet those goals:

> (A) building resiliency against climate change-related disasters, such as extreme weather, including by leveraging funding and providing investments for community-defined projects and strategies;

(B) repairing and upgrading the infrastructure in the United States, including—(i) by eliminating pollution and greenhouse gas emissions as much as technologically feasible; (ii) by guaranteeing universal access to clean water; (iii) by reducing the risks posed by climate impacts; and (iv) by ensuring that any infrastructure bill considered by Congress addresses climate change;

(C) meeting 100 percent of the power demand in the United States through clean, renewable, and zero-emission energy sources, including— (i) by dramatically expanding and upgrading renewable power sources; and (ii) by deploying new capacity;

(D) building or upgrading to energy-efficient, distributed, and "smart" power grids, and ensuring affordable access to electricity;

(E) upgrading all existing buildings in the United States and building new buildings to achieve maximum energy efficiency, water efficiency, safety, affordability, comfort, and durability, including through electrification;

(F) spurring massive growth in clean manufacturing in the United States and removing pollution and greenhouse gas emissions from manufacturing and industry as much as is technologically feasible, including by expanding renewable energy manufacturing and investing in existing manufacturing and industry;

(G) working collaboratively with farmers and ranchers in the United States to remove pollution and greenhouse gas emissions from the agricultural sector as much as is technologically feasible;

(H) overhauling transportation systems in the United States to remove pollution and greenhouse gas emissions from the transportation sector as much as is technologically feasible, including through investment in—(i) zero-emission vehicle infrastructure and manufacturing; (ii) clean, affordable, and accessible public transit; and (iii) high-speed rail;

(I) mitigating and managing the long-term adverse health, economic, and other effects of pollution and climate change, including by providing funding for community-defined projects and strategies;

(J) removing greenhouse gases from the atmosphere and reducing pollution by restoring natural ecosystems through proven low-tech solutions that increase soil carbon storage, such as land preservation and afforestation;

(K) restoring and protecting threatened, endangered, and fragile ecosystems through locally appropriate and science-based projects that enhance biodiversity and support climate resiliency;

(L) cleaning up existing hazardous waste and abandoned sites, ensuring economic development and sustainability on those sites;

(M) identifying other emission and pollution sources and creating solutions to remove them; and

(N) promoting the international exchange of technology, expertise, products, funding, and services, with the aim of making the United States the international leader on climate action, and to help other countries achieve a Green New Deal.

Notice that this list covers the four pillars of deep decarbonization: improved energy use efficiency; electrification of vehicles and nearly all gadgets and industries that currently use fossil fuels; using renewables as the main source of electricity; and R&D on carbon capture and storage technologies. We could debate some details, such as whether it will be economically feasible to reach 100% renewables by 2050, or whether there will still be a need for 10%–20% of electricity generation fueled by natural gas or nuclear power to stabilize electricity supply, with the 100% goal achieved in the following decades. The answer to these and other questions, like powering aviation, steel manufacturing, maritime shipping, and emissions from cement making, will depend largely on technological developments that are difficult to predict three decades in advance. The GND resolution goes beyond deep decarbonization of the energy, manufacturing, and transportation sectors to also include agriculture, forest conservation and regeneration, and foreign aid for transfer of technology, knowledge, and services.

More important for us to recognize here is that the GND resolution and the four pillars of deep decarbonization are mostly silent on the approaches for achieving their goals and carrying out their initiatives. Searching for words like "tax" and "regulation" in this document yields nothing. So, how do we get there from here? This chapter has presented a few somewhat more specific policy approaches, although still with many details to be determined by policy makers and the public:

Put a price on carbon while buffering the impact on low-income groups. This chapter discussed the pros and cons of a carbon tax with dividends, cap-and-trade systems, incentives and fines, and other approaches using regulations and tax incentives. I argue for using all of the tools in the toolbox, each strategically selected as the right tool for the specific job. However it is done, putting a price on carbon will incentivize consumers, corporate executives, engineers, farmers, and policy makers to pursue the myriad intelligent choices needed to reduce reliance on fossil fuels and to develop renewables. I have already argued forcefully that money is not everything (trees and bees are important too), but I also respect that the pocketbook is a remarkably potent motivating force. Let the force be with us, not against us. The marketplace, for all of its shortcomings, can also be an effective tool, given proper guardrails, to advance needed technological innovation. In addition to listening to economists and environmental scientists, we must also listen to social scientists and the people who will be paying the price put on carbon. Some or all of the revenues raised should go back to those who will have difficulty paying the temporarily increased prices on energy or to families of those who lose their employment as the fossil fuel industry shrinks.

Increase incentives for entrepreneurs and early adopters of energy-efficient and renewables-based technologies. Many forms of solar and wind energy production are now competitive with fossil fuels without subsidies, but subsidies for producers and rebates and tax deductions for early adopters were crucial for accelerating the pace of achieving their economic competitiveness. The same is urgently needed now for electric vehicles, battery storage, heat pumps for heating and cooling homes and buildings, and other innovations too numerous to list or not yet invented. *Current enormous government subsidies for fossil fuel use should be eliminated or shifted to renewable generation of electricity for all sectors of the economy.*

Expand R&D on the technologies that will be most difficult to electrify and to run with renewable energy, such as long-haul trucking, maritime shipping, aviation, steel manufacturing, and cement making. Acknowledging that some fossil fuel use will be extremely difficult to eliminate soon, *we need R&D on making carbon-capture-and-storage technologies more economically viable* so that they can be used to offset remaining emissions from fossil fuel burning. Beyond fossil fuels, carbon capture and storage also holds promise as a means to draw down atmospheric CO_2 toward the latter half of this century, putting us back on a path to pre–industrial revolution atmospheric CO_2 concentrations and climate. Finally, *R&D on effective and economical battery storage technology is urgently needed,* because it may be a bottleneck for moving as quickly as desirable to nearly 100% renewables.

Support the regenerative agriculture policies recommended in the previous chapter, because they will also help reduce greenhouse gas emissions.

Beyond decarbonizing our economy, we need to leverage the large role that forests play in the global carbon cycle and climate change. We should *support international efforts to avoid adding CO_2 to the atmosphere through tropical deforestation and promote removing CO_2 from the atmosphere through reforestation of degraded lands.* These include initiatives such as: encouraging private sector supply chains of agricultural commodities to avoid products grown on recently deforested land; funding international programs like Reduced Emissions from Deforestation and Degradation, carrying out the goal of designating 30% of the landscape for conservation in native vegetation by 2030; supporting the protection of indigenous lands; working with local experts and stakeholders to find effective and socioeconomically desirable ways to reforest degraded lands; and helping educate and train the human capital needed in developing countries to carry out their own initiatives and law enforcement to sustainably manage their natural resources.

Create a Civilian Climate Corps. This idea is borrowed from Franklin Roosevelt's New Deal Civilian Conservation Corp (CCC), which employed about 3 million men (segregated into camps by race) on environmental projects during the Great Depression. The CCC constructed trails and shelters that shaped today's modern national and state park systems. A modern CCC would similarly provide person power (young men and women from diverse communities) to advance energy conservation, renewable energy installation, and reforestation and forest thinning (as fire-prevention) efforts. My life was changed by a Peace Corps experience as a young adult, so this proposal is near and dear to my heart. America would reap many benefits of climate change mitigation and adaptation projects from this initiative, but perhaps the most important impact would be on the lives of the young participants, who would learn from valuable hands-on experiences.

Hold countries and the private sector accountable to their COP26 pledges. The pledges made in Glasgow in November 2021—to reduce methane emissions 30% by 2030, to stop deforestation by 2030, and to commit $130 trillion to transform the economy—are considerably more ambitious than pledges from previous COP meetings, although still not ambitious enough to keep the global mean temperature rise below 1.5 degrees Celsius.[73] However, past, present, and future pledges are not worth the paper they are printed on unless they are honored with real actions and real money. The Paris Climate Accords and subsequent COP agreements depend on countries living up to their voluntary, nationally determined contributions to emissions reductions, which, in turn, means that citizens must hold their national leaders accountable.

6

The Luddites Had It Half-Right, but the Other Half Could Be Great News

British weavers and textile workers in the late eighteenth century had good reason to fear an existential threat from the mechanization of their industry. Knitting machines made possible by the emerging industrial revolution could make products a hundred times faster than those made manually by skilled craftsman and textile artisans. The machines could be run by fewer and less skilled laborers who were paid lower wages. In 1779, a short-lived rebellion of self-proclaimed "Luddites" was purportedly led by a General Ludd, who, like the mythical Robin Hood, supposedly lived in the Sherwood Forest. The Luddites attempted to smash machines in the newly emerging textile factories, but the rebellion was put down by British soldiers. A century or more later, the word *Luddite* was resurrected in the popular lexicon as a derogatory term for anyone who dislikes new technology or who blindly opposes technological progress. As a child experiencing color television for the first time, I equated anyone who did not like that marvelous new technology with being a stupid Luddite.

The Luddites of 1779 saw the stitching on the wall, that they were about to be replaced by machines, which, nowadays, also includes robots (here I use the words *machine* and *robot* almost interchangeably, because robots that use data to determine how to perform functions are embedded within many modern machines; they do not need to look or act like the endearing R2-D2 to be called a robot). Following their unsuccessful rebellion, some Luddites were executed, others were sent off to Australia as prisoners, and the remainder saw their jobs disappear. Today a handwoven garment may fetch a premium price, but that kind of artisanship is a niche profession that employs relatively few people.

Would humanity have been better off if weaving machines had never been invented? Would we be better off without tractors, automobiles, televisions, or cell phones? In one respect, these are pointless questions, because human ingenuity is always developing new technology. Innovation is part of being human, leading our prehistoric ancestors to learn to tame fire, domesticate wild plants as crops and wild animals as livestock and pets, and smelt metals into useful forms. On the other hand, humans also have the introspective capacity to examine the unintended consequences of our new technologies and to consider how to avoid or mitigate their undesirable impacts while enjoying the potential benefits, efficiencies, and safeties that they offer.

Only recently has human society encountered technologies that pose truly existential threats and that require very serious restraint for the safety of all humanity. Weapons of war that cause misery and death date back to our prehistoric ancestors, but modern nuclear weapons pose a new threat in their capacity for global annihilation, and so we must now absolutely control their proliferation and prevent their use. New ethical challenges are arising as biomedical technology enables editing of the human genome, thus putting into question what it means to be a human. Artificial intelligence is greatly enhancing our ability to make sense of huge datasets and engineer new designs and products, including new medicines, but we do not yet fully grasp its potential for humanlike cognition and the possible multiple impacts on social and political values and ethics.

In this chapter, we focus on ongoing technological advances that are replacing human workers. As we saw in the previous chapter on climate change, the impacts of Anthropocene technologies, such as fossil fuel combustion and ozone-eating chemicals, have created existential threats to a habitable planet. Avoiding those threats will require a profound technological change in the way we harness energy for the economy. These changes will inevitably cause disruptions, with some winners and some losers, at least in the short term.

The Robot in the Coal Mine

Few of us would assert that weaving machines, farm machinery, automated teller machines, word processing, and robots that assist with coal mining and automobile assembly are inherently unethical. Yes, they have replaced human workers, but the remaining workers in those industries become more productive in terms of economic output per worker, and they are probably safer, too. Miners used to carry caged canaries into coal mines as indicators of unsafe concentrations of methane or carbon monoxide in the air; if the canaries died, then the miners were warned to get out. Machines that monitor the air have replaced the canaries, and now robots and other machines are replacing the miners.

There is no doubt that technological innovations cause social disruption, good and bad, and the Luddite weavers are only one historical example. Coal mining is declining today because of a double whammy from two new technologies—coal mining mechanization and natural gas from fracking—resulting in a precipitous drop in the demand for coal, closing of coal mines, and the loss of jobs for thousands of coal miners. Modern mechanization of coal mining began on a large scale in the United States in the 1950s (long before climate change became an environmental and then political issue). Miners have systematically been replaced by mining machines, just as the Luddites were replaced by weaving machines nearly two centuries earlier. Mountain-top mining, which became common in the 1980s and 1990s, also used less labor. As a consequence of improved machinery and growing demand, US coal production doubled from 1950 to 2000, but the number of human miners dropped from 470,000 to 80,000 and to less than 50,000 today (fig. 6.1). Because of mechanization, each coal miner became about 12 to 14 times more productive as

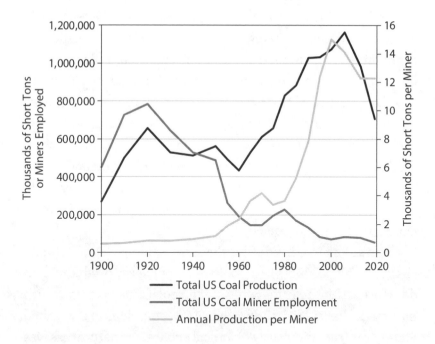

FIGURE 6.1 US Coal Production, Employment, and Worker Productivity. Drawn by the author from data provided in Global Energy Monitor Wiki and from the US Energy Information Administration, *Annual Coal Report,* October 5, 2020, https://www.eia.gov/coal/annual/

measured by coal extracted per human miner, but there are many fewer jobs in coal mining.

This trend of increasing worker productivity does not necessarily translate into higher wages for the remaining mine workers. To the extent that the remaining workers are more skilled, such as being trained to operate and maintain high-tech robotic machines, some of them may receive higher pay. However, most of the financial benefit of higher worker productivity is enjoyed by the employer. As discussed in chapter 2, unions, including those that represent miners, have lost power during the last 40 years, which has contributed to the growing gap in wealth and income between the middle classes and the wealthiest 10% (recall fig. 2.1).

Despite the increase in productivity that mechanization brought the coal industry, the discovery of a way to extract natural gas by injecting pressurized liquid into deep layers of rock, called hydraulic fracturing—*fracking* for short—has made coal economically uncompetitive in many markets today. Consequently, the remaining miners not yet replaced by machines now face closure of their mines because additional technological developments made natural gas more economically attractive to the utilities, industries, and consumers that can substitute it for coal. The COVID-19 pandemic further reduced demand for energy, at least temporarily, causing some of the more marginally profitable coal-fired plants and coal mines to close, and once closed, they may not reopen.

Natural gas is a fossil fuel that still contributes to climate change, but it produces less CO_2 per unit of energy produced than does coal. Although the widespread replacement of coal with natural gas has significantly reduced CO_2 emissions in the United States, this transition away from coal and toward natural gas was driven not so much by environmental policy, but rather mostly by the economics of supply and demand of a cheaper product enabled by technological development. The current decline in coal mining is not, as some have falsely claimed, because of a war on coal, any more than hand weaving declined because of a war on Luddite weavers. Some environmentalists may be cheering on the sidelines about the decline of coal because it means lower CO_2 emissions, but it would be inaccurate to say that they were the driving cause of coal's decline. Weaving machines were invented because they made clothes faster and cheaper, not because of a politically driven war on weavers, and coal has been in decline primarily because technology has left the coal industry behind. Any politician who says otherwise about the causes for the trends to date illustrated in figure 6.1 is not being honest. That said, further acceleration of the current trend of declining coal production will be necessary to meet deep decarbonization goals, which is needed for both environmental and economic reasons.

Coal miners correctly perceive a threat to their livelihoods, but they are not Luddites in the modern, pejorative, antitechnology sense of the label. Nevertheless, unless they can adopt other professions or unless society can provide them with assistance during this transition, their fate could be similar to the Luddites (presumably without violence). In the meantime, the remaining coal miners have become more productive and the industry, with more robots and fewer human workers, has a somewhat better fighting chance of remaining economically competitive (for example, a class of coal called *metallurgic coal* is still needed for steel making). A silver lining that merits emphasis is that the robots do not suffer from black lung disease, and fewer humans are put in harm's way in the event of serious mining accidents, which are historical tragedies of the industry.

How Many Farmers Does It Take to Feed Us?

The biggest labor transition in human history was and still is from an agrarian rural society to an industrial, mostly urban society. While mostly complete in North America and Europe, that transition is still ongoing in many parts of Asia, Africa, and South

America, which are home both to megacities and to still significant rural populations. All of the able-bodied adult women and most of the men in the African village where I lived went to their fields every day except Sunday, harvesting from their small farm plots nearly everything that they would eat each day. The vast majority of the children in the village were expected to do the same when they became adults, although that is changing, even in rural Africa.

Before the industrial revolution, the majority of people in all countries were farmers—about 90% in the United States in 1800. Today, on-farm jobs make up less than 2% of the work force in the US, with similar percentages in other high-income countries. On the downside, most small rural towns have suffered declines in population and income. Those that have survived and prospered have diversified beyond their agronomic roots to become centers for small industry, health care, education, and tourism. In the meantime, thanks to the hard work of the relatively small number of remaining farmers, the rest of us are free to pursue other endeavors, without worrying about working the fields every day for our daily meals. The technology that allows the average American farmer each to feed well over one hundred people now also enables us to devote labor and innovation to other sectors, from education to industry to the arts.

When people moved out of the farming sector, starting in the nineteenth century and continuing today, what did they do for employment? Most of the descendants of those farmers found jobs in factories or in the many service sectors in cities, from finance to real estate and from energy to education. As machines and robots now replace more and more factory workers and service-sector workers, making those sectors more efficient in terms of labor, energy, and reduced greenhouse gas emissions, where will those jobs go? Some worry about rising unemployment, which is a legitimate concern, as we are currently seeing in economically depressed coal country. But just as we now think it normal for farmers to make up only a tiny percentage of the developed world's workforce, so, too, may we one day think it normal for robots to do the majority of work in mines and factories.

In the long view, the transition from rural agrarian to urban industrial economies has mostly worked because industry and service-sector jobs were created in urban centers. The steam engine, the weaving machines, and the harnessing of fossil fuels in industry and on the farm unleashed an explosion of productive capacity. Notwithstanding the pollution and labor abuse that accompanied it, this urban-based industrialization led to a general long-term increase in economic productivity, employment, and material wealth. While wealth disparities remain huge, on average, we are better nourished, better educated, and have longer life spans and higher standards of living than the people living before and during the early stages of the industrial revolution.[1] As discussed below, however, there is no guarantee that new job growth in other sectors will replace the jobs now being lost due to ongoing automation and other technological developments.

As a natural scientist, I return to the basics for what generates wealth in the first place—our skill and ingenuity in extracting essential natural resources that we transform to provision humanity with food, water, shelter, clothing, energy, and materials for industry. Those transformations of resources to fulfill human needs then also enable further transformations of human capital to produce additional knowledge, skills, and services that create more jobs, wealth, and well-being. The Earth's natural resources themselves are finite, but the ideas and ingenuity for sustainably harnessing and recycling those resources may be infinite and certainly have great potential for augmenting human well-being. When technological innovations, including but not limited to automation, increase our capacity to sustainably manage those natural resources to produce more nutritious food, reliable supplies of potable water, raw materials for construction and industry, and renewable energy, then more wealth can be generated to support human endeavors of all kinds. We need not all be farmers or miners. Most importantly, following the advent of technological developments, the number of farmers and miners no longer primarily determines the production of food and energy. When we have abundant, affordable, and sustainably produced food and energy, then human ingenuity is unleashed to produce multiplicative layers of goods and services, including meaningful employment along the way. In theory, then, a small fraction of the labor force supervising robots and machines should be able to support many times as many people providing refined goods and services to each other.

Sustainable exploitation of natural resources generates the basis for wealth, and then technology, innovation, and labor further amplify that wealth up the food chains, supply chains, technology chains, and knowledge chains, thus creating respectable jobs and the goods and services that we enjoy consuming. I am not arguing for unbridled or indefinite growth of consumption or for using only economic metrics like gross domestic product. Rather, I argue for a clear recognition of the bounty that we derive from the Earth's natural resources and the role of appropriate technology to sustainably steward those resources and to multiply the wealth and well-being that they provide. Moreover, this vision for emphasizing the positive aspects of technological transitions is not inevitable and will not just happen spontaneously, as my friend Bruce from chapter 1 might have assumed. Nor is it due to any perceived "natural law" of the marketplace. Indeed, avoiding unemployment as an unintended consequence of automation will require some innovative green new deal thinking, planning, and policy.

Can Renewables and the GND Bring Back Better Jobs?

Automation is happening throughout our economy, so the question of employment opportunity extends far beyond just the coal industry or the agricultural sector. An unintended consequence of the new wave of technological development may well be another shift in the workforce similar in scope to the transition in Europe and North America from an agrarian society prior to the industrial revolution to the industrialized society of the late twentieth century. We don't know how this will play out, and there is notable disagreement among economists on this subject. On the one hand, automation reduces employment, and the resulting surplus of workers looking for scarce jobs can lead to lower wages if new jobs are not created elsewhere in the economy. This seems to have been the case in the late eighteenth and early nineteenth centuries, when the Luddites and their other artisan contemporaries lost their jobs as the industrial revolution progressed; wages were depressed for decades during that era and poverty was widespread (think of Charles Dickens's Bob Cratchit and Tiny Tim).

On the other hand, when automation leads to greater efficiency, productivity, and profitability, and *if* those profits are invested in the development of new industries and services to meet the increased demands of those who benefited from greater efficiencies, then new jobs are created. This happened eventually in nineteenth-century England, with an increased demand for machine-woven fabrics and thus an increased demand for yarn and associated jobs to feed the profitable mechanized weaving industry.[2] However, it took decades before employment and wages recovered from this transition in the industries affected by the industrial revolution. It was not the generation of Luddites and other contemporary artisans who benefited from industrialization, but rather their children or, more likely, their grandchildren. Likewise, the children and grandchildren of farmers who adopted labor-saving agricultural technologies have, for the most part, been absorbed by jobs found in the city, albeit no doubt with many personal hardships and achievements along the way.

In addition to generational transitions, technology can cause geographic disruptions. When technological changes reduce employment in one geographic region while favoring employment using a technology in another region, such as the substitution of coal with natural gas from fracking, the unemployed workers from one region often do not immediately move to follow the jobs, as economic theory might suggest rational actors should do. Instead, a large fraction of the laid-off workers are reluctant to leave family and friends and the cultural heritage and landscape that they know, however economically irrational that may seem to theoretical economists. Relocating for uncertain opportunities in an unknown and seemingly risky new endeavor is often too steep of an impediment, and maybe that is rational after all. Economists refer to these human social and emotional aspects that interfere with realization of market efficiencies as *stickiness*.[3] Only if their condition becomes so dire that they face near starvation do large numbers tend to migrate, as was the case for the Dust Bowl farmers of Oklahoma during the Great Depression or the case of modern environmental refugees discussed in chapter 3. The current case of coal mining disruption is severe but not yet quite as cataclysmic. Already, however, young people are leaving those communities to seek education and employment opportunities elsewhere, and they are unlikely to return. Economic theory may be mostly right in this case, albeit with a notable lag until a fraction of the next generation seeks out opportunities away from home (being less sticky), while most of their parents and grandparents choose to persevere in the depressed economy of their homeland.

While the examples of the industrial revolution of the eighteenth and nineteenth centuries and the rural-to-urban transition following technological innovations in agriculture are instructive, they are not perfect analogs to the automations that we are experiencing today. Economists are less certain how our current wave of automation—which is replacing workers in toll booths, grocery checkout stands, coal mines, and automobile manufacturing plants, to name a few—will result in efficiencies and profitabilities that could stimulate growth and new jobs in new industries and services. To be sure, the computer age has created jobs that never existed before, such as web designers, professional gamers, and Instagram influencers, and it has transformed existing fields like engineering and education. The proliferation of renewable energy will also create new job opportunities.[4]

Some economists favor instituting a robot tax on businesses while reducing payroll taxes on employees, which would help level the playing field of robots versus human workers. South Korea has done so, but most countries are reluctant to interfere with technological progress exemplified by automation. Research and experimentation are needed to find the right balance of technological development and human employment and to mitigate the hardships experienced by generations of workers in regions that undergo technological transformations.

We don't yet know whether the creation of new jobs will keep up with the loss of old jobs due to current trends of increasing automation, but we know that employment policy must be part of any green new deal initiative, whether it be advancing renewable energy, electric car and battery manufacturing, or R&D on cement and steel manufacturing. Indeed, a transition guided by green new deal thinking offers a potential solution to the employment conundrum presented by market-driven and technology-driven automation.

Access to Education

Another key to transforming jobs along with technology is education. We don't know what happened to the children of the Luddites, but we do know that very few of them had access to more than the most elementary education, and it is reasonable to speculate that they didn't prosper. Sadly, the industrial revolution created conditions ripe for abusive child labor. When I toured the jute factory in Dundee, Scotland, where my great-grandfather worked in the late nineteenth century, I learned that children were employed because their small bodies and small hands could reach into tight places to repair machines, but with great risk to life, limbs, and fingers. There were many more women employed in the factory than men because they could be paid less and had greater tolerance for the repetitive machine work. A generation of men who couldn't find work stayed home to take care of children while their wives worked, until World War I created a calling, and frequently a tragic end, for those unemployed men. Child labor is now mostly a thing of the past in the developed world; the gender gaps for wages and opportunity have narrowed but not yet closed; and most, but clearly not all, descendants of those early industrial revolution workers have access to varying levels of education. I enjoyed that privilege.

Where I now live in Appalachia, two- and four-year colleges have recently started working with local businesses to figure out their workforce needs. Based on that information, these colleges are developing curricula for training the next generation of workers, mostly from the local area, to meet those needs. Students and their families still struggle to pay tuition and room and board, but one of our local counties has made its two-year community college tuition-free for local residents. If rural Garrett County, with its modest financial resources, in the far-western tip of Mountain Maryland can make community college tuition-free, surely a green new deal can generate a coalition of county, state, and federal governments to make all two-year colleges tuition-free. With that done, we can aim for other financial assistance for those students and to make both two-year and four-year colleges affordable for all and without barriers for minoritized communities.

While readin', 'ritin', an' 'rithmetic remain basic skills for educational goals, advanced education is preparing students for jobs never dreamed of by the originators of the three Rs. Designing video games and cybersecurity are good examples of jobs that never existed before the digital revolution of the late twentieth and early twenty-first centuries. Our grandparents would have had no idea what it meant. Indeed, when I told my grandfather that I was going to go to graduate school in forestry, he was not impressed, but the meaning of that named profession had changed since his day. He envisioned that I would spend my days in a lookout tower, like Ranger Rick, watching for forest fires, when in fact management of forest resources is very complex in ways that my grandfather had never imagined. Future generations will most likely be managing new ways of growing food and fiber, producing and distributing energy, and creating products and services that we have not yet imagined.

Education is necessary not only to train people who can fill existing jobs and new ones as they come on line, but also to enable the innovators who will create the industries and markets that produce those jobs. Such inclusive investments in human potential must be central to a green new deal. The result will be a proliferation of diverse innovators who create new technologies, products, services, and jobs, while leaving fewer people behind.

A better-educated workforce has a fighting chance to demand a larger piece of the pie of wealth that they help generate, but it is not a guarantee that the distribution of wealth will become more equitable. The Luddites predated the labor movement; the coal miners had strong unions and good wages for a few generations, but then lost much of that power. It remains to be seen whether workers in new industries will gain bargaining power through the traditional labor union approach, through new mechanisms not yet invented, or remain as vulnerable as the gig-economy Uber drivers are today. Obstacles to union organizing are generously funded by the wealthy and influential individuals who benefited from declines in the strength of organized labor during the last four decades. On the other hand, many workers are refusing low-paying jobs as the economy recovers from the pandemic, requiring many employers to offer higher wages to attract workers. It's too early to know how this fluid situation will pan out, but a likely challenge of green new deal thinking will be to find common ground among various stakeholders to fill labor shortages while ensuring fair compensation at all levels of employment.

Technological increases in efficiencies, if done correctly, could also extend the historical trend of shortening the workweek. Whether a shortened workweek will mean less weekly pay remains to be seen; that is not what has happened historically, but it could happen during the new wave of automation. Employers are experimenting with expecting greater worker productivity in fewer hours of work per week, which would allow the same or greater pay for a shorter workweek. Work may become less hostile to family needs as flexible hours and telecommuting become more feasible and culturally acceptable, which the explosion of the Zoomosphere during the pandemic has already demonstrated is possible. In any case, the nonmonetary benefits of greater worker efficiencies and flexibilities could have welcome consequences, such as providing more time for parents to spend with their kids, care for their elderly parents, pursue interests outside of work, and relieve stress. Greater worker productivity could also mean that we need fewer younger people to support a growing cohort of retirees, thus allowing population growth and its attendant pressures on natural resource demands to level off or even reverse.

Pathways to Smart Diversified Employment

In the Appalachian region of western Maryland where I live, the local communities are working hard to attract small industries to the region to make up for lost employment in the coal industry and paper mills. We have a lot to offer—a relatively low cost of living, a beautiful mountainous landscape, a friendly and safe community, little traffic, and many recreational activities. I attended monthly meetings of a community group that is promoting internet expansion, improved highway networks, and an upgrade of our local airport to broaden the appeal of our community for businesses and families to locate here. Encouraging greater use of highways and airplanes is not consistent with the long-term approaches espoused by a green new deal, but I understand my community's eagerness to use the technology and opportunities at hand to thrust ahead now.

At the same time, the region is succeeding in attracting some new technologies. A start-up company is developing technologies to extract valuable rare earth metals (a subject of chapter 7) from old mine spoils. A solar panel farm is going ahead on land that was only minimally reclaimed after surface mining and thus not productive for agriculture or forestry. Natural regrowth of a forest has been extremely slow at that site due to the poorly reclaimed soil conditions. Meanwhile, the local county commissioners are delighted, because the solar farm will generate new tax revenue and jobs from otherwise idle land.

Rural Internetization Is to a GND as Rural Electrification Was to Roosevelt's New Deal

The internet is a virtual highway, which, among other things, enables workers and their families to enjoy living in rural communities while telecommuting to work most days. I know people with well-paying jobs who physically travel to a regional headquarters office only once or twice a month. The COVID-19 pandemic has ushered in an unprecedented acceptance of telecommuting that is likely to stick, at least in part, opening up new possibilities for rural communities to attract new telecommuting residents, injecting new revenues into those communities. This is only possible, however, where there is good internet connectivity, which is too often not the case.

Just as the Rural Electrification Act of 1936 was part of President Roosevelt's New Deal to help bring prosperity to rural America, we are overdue for a green new deal that would bring high-speed internet to rural communities throughout the world. At present, private internet providers have trouble making a profit in sparsely populated areas, which economists call a "market failure."[5] When the free market fails to provide an essential service, some government subsidies or incentives are needed to make it economically viable. The investment would not only open doors to telecommuting, it would also provide opportunities for adults who have lost jobs in rural areas to seek on-line retraining. The pandemic demonstrated that telemedicine can be a valuable addition to rural communities, but it, too, requires that residents have access to reliable internet service. Students in these rural areas were also handicapped during the pandemic by lack of good access to the internet for distance learning while their schools restricted in-person classes (students in low-income urban areas also often lack internet access, not because it doesn't exist, but because it isn't affordable).

While many displaced workers from the coal industry and other sectors affected by automation are unable to overcome the many socio-economic impediments to transitioning to other employment, some of those workers have taken advantage of retraining opportunities, enabling them to benefit from new jobs as other sectors of employment grow. The US Department of Labor has a small program called the Trade Adjustment Assistance (TAA) program, which provides a variety of benefits and reemployment services to workers who have lost their jobs. When a local paper mill in my community suddenly closed because its main product could be produced more efficiently elsewhere, the TAA program enabled several laid-off workers to enter a retraining program at the local community college, while also receiving a stipend that kept food on the table for their families. The TAA program started before anyone coined the term green new deal, but it exemplifies the same convergence of social science with education and employment policies. Many communities facing high unemployment rates due to layoffs and transitioning technologies do not know about this program. Even with the best policies and intentions, however, technology-driven changes in employment will still be disruptive, necessitating health care and financial assistance for workers like former coal miners and their families for at least a generation.

Another long-range investment in US rural communities would be to revitalize our rail system. A major freight railroad company is still an important employer in my community, but the employment trajectory is downward. Amtrak, the US passenger train, comes through once a day, but it often runs late and is not a viable option for routine traditional commuting to the nearest metropolitan areas of Washington, DC, Baltimore, or Pittsburgh. The US rail system is generally an embarrassment compared to those in China, Japan, and most of Europe. The explanation is fairly straightforward—the US federal and state governments have subsidized construction and maintenance of highways more than railways ever since World War II, spending hundreds of billions of tax dollars on roads since the 1950s. Government investment in roadway technology ensured that cars and trucks would become America's primary means of transportation for passengers and even for freight. It should not be surprising, then, that railways have declined in importance while the trucking industry has grown.

These disparities are not the result of Adam Smith's invisible hand of capitalism, but rather the direct response of policies to subsidize highways. A green new deal could equalize or reverse these incentives and subsidies so that entrepreneurs and technological innovators would find it worth their investments in time and money to upgrade a mostly antiquated rail system. Rural communities would benefit, shipping and transportation would become more energy efficient, and greenhouse gas emissions would decline. Trucking would not go away, but rail would become more competitive for freight. On the other hand, the advent of electric trucks fueled by renewable energy, as discussed in the last chapter, could again shift the balance between trucking and train freight in unexpected ways. The success of the Acela train service on the eastern corridor of the United States demonstrates that many Americans, like most people around the world, would use a relatively fast and reliable train service if it were available, potentially reducing demand for aviation fuel, for which we do not yet have an economically viable green technology substitute.

The Luddites of the Fourth Industrial Revolution

The Luddites got a bad rap from historians. They were not stupid; on the contrary, they were very discerning in recognizing that machines were going to disrupt their livelihoods. Since then, we have muddled through the industrial revolution and the twentieth century, with both advances and disruptions, *but* are we satisfied with the tragic fate of the Luddites and the many other workers in jobs that have since become obsolete? If we are to do more than stand by as more and more coal miners and other displaced workers meet the same fate, we will need new transition policies that integrate technology, employment, the environment, and social conditions.

A vision of the fourth industrial revolution casts new technologies into a social context.[6] According to this account of history, the first industrial revolution was the era that devastated the Luddites while unleashing the power of steam generated by burning fossil fuels to run machines, boats, and trains; the second saw widespread electrification, which preceded and followed the New Deal of the 1930s; the third used electronics, computers, and information technologies, which pervade our lives today; and now the fourth grows out of the digital revolution begun in the third. Artificial intelligence is one example of the accelerating power of the fourth industrial revolution, which is already being used for both bad (for example, digital facial recognition to spy on citizens by authoritarian governments) and good (improved medical diagnoses and other advances in science; improvements in engineering such as designing a circular economy of efficiently distributing and recycling materials, which is the topic of the next chapter). Without guidance, the bad looms ominously as an Orwellian enabler of authoritarian repression, and muddling through this new era could very well land us all there. As part of a green new deal, however, we will need to anticipate and proactively hold in check the unintended and undesired social implications of the new technologies of the fourth industrial revolution, while the positive outcomes enable profound, convergent pathways toward environmental, economic, and social sustainability.

The examples from the past and present discussed in this chapter demonstrate how technological developments have had so many unintended consequences on peoples' livelihoods—from eighteenth-century Luddites to modern coal miners. *But*, have we ever tried to foresee those consequences and manage them to maximize the benefits of technology, while minimizing the negative impacts? While clearly disruptive and potentially fraught with both foreseeable and hidden perils, technology offers hope to solve some of our most wicked problems, including climate change and pandemics, provided that it converges with green new deal thinking about the resulting social, economic, and environmental consequences.

Recommendations

The Green New Deal (H. Res. 109, 116th Cong., 1st Sess. [introduced February 7, 2019]) has much to say about sources and qualities of future employment. The proposal emphasizes new jobs in the renewable energy sector, but it also calls for investing in existing manufacturing and industry and assisting where transitions away from old industries will be challenging:

> spurring massive growth in clean manufacturing in the United States and removing pollution and greenhouse gas emissions from manufacturing and industry as much as is technologically feasible, including by expanding renewable energy manufacturing and investing in existing manufacturing and industry; . . .

> directing investments to spur economic development, deepen and diversify industry and business in local and regional economies, and build wealth and community ownership, while prioritizing high-quality job creation and economic, social, and environmental benefits in frontline and vulnerable communities, and deindustrialized communities, that may otherwise struggle with the transition away from greenhouse gas intensive industries.

It also emphasizes working conditions and a living wage:

> a Green New Deal must be developed through transparent and inclusive consultation, collaboration, and partnership with frontline and vulnerable communities, labor unions, worker cooperatives, civil society groups, academia, and businesses.

Promoting education to enable the next generation of workers is also emphasized:

> providing resources, training, and high-quality education, including higher education, to all people of the United States, with a focus on frontline and vulnerable communities, so that all people of the United States may be full and equal participants in the Green New Deal mobilization; . . .

> ensuring that the Green New Deal mobilization creates high-quality union jobs that pay prevailing wages, hires local workers, offers training and advancement opportunities, and guarantees wage and benefit parity for workers affected by the transition.

Green new deal policies will no doubt create many new jobs in renewable energy, energy efficiency, regenerative agriculture, and innovative technologies in many sectors. However, other jobs will inevitably be lost as old technologies become obsolete, whether that obsolescence is caused by ongoing automation that has nothing to do with a green new deal or is accelerated by replacement technologies encouraged by a green new deal. Regardless of the causes, a green new deal can help with the solutions. The following list therefore emphasizes a convergence of community-driven engagement with social scientists, economists, and policy experts to help ensure that both the newly employed and newly unemployed are fairly and justly treated and to bolster the vitality of rural communities:

Provide resources for rural communities to develop their own locally specific economic and cultural development initiatives and connect them with networks of communities to exchange experiences and the knowledge that they co-produce.

Expand access to high-speed internet throughout the country and the world, subsidizing service to rural communities if necessary. This may be the most effective means of bolstering the economies of rural communities that lose their historical extractive industries. It would also enable improvements in the quality of education and health care for children and adults in those communities.

Make two-year colleges tuition-free in the United States and set goals for making four-year colleges and universities more affordable. Incentivize colleges and local businesses to collaborate on aligning curricula with locally and regionally needed workforce skills. Support historically Black, Hispanic, and Indigenous colleges and universities that provide additional supportive educational environments for minoritized communities. Support similar expansion of affordable educational opportunities for both boys and girls throughout the world.

Modernize freight and passenger rail service, including access to rural communities.

Support research to identify and mitigate the downsides of the emerging technologies of the third and fourth industrial revolutions, such as infringements of personal liberties and loss of employment, while fostering the upside benefits to humanity.

Expand the US Department of Labor's Trade Adjustment Assistance program, and similar federal, state, and local programs, which provide a variety of benefits, training, and reemployment services to workers who have lost their jobs. Incentivize local two- and four-year colleges to coordinate with these programs.

Enact policies that ensure food security and access to health care for those who have lost jobs due to shifts in industries. These people should not feel forgotten or abandoned.

7

There's a Great Future in ~~Plastics~~ <u>Circular Economies</u>

Dustin Hoffman's young character in the 1967 film *The Graduate* is pulled aside by an older businessman and admonished to think of the future in one word: *plastics!* At that time, the global production of new plastics was about 15 million tons. The advice was prescient, as plastic production grew to over 300 million tons in the following 50 years.[1] That may have been good advice for a 1960s business opportunity, and many useful products have been created, but it hasn't turned out so well for the environment, and it may be bad for human health. Plastic is nearly ubiquitous in the products we purchase; we are awash in plastic waste; and tiny plastic particles are fouling our air, water, and food. We are just beginning to understand the scope of its environmental and human health consequences.

For all of the waste and sustainability issues discussed so far in this book—from manure to greenhouse gas emissions to plastics—I would have a much different admonishment for today's young, enterprising businessperson: think of *two* words for the future: *circular economy!* It is both a good business opportunity and a win-win for the environment and human health.

A plastic bag that is designed to be used once and thrown away will most likely fulfill that designed role and nothing more—nothing good, anyway. Some conscientious do-gooders will put their used plastic bags in a recycling bin designated for that purpose if they can find one, others will at least throw them in a proper garbage can, but far too many will cast them aside to become ugly and dangerous litter across landscapes, rivers, lakes, and oceans. Plastic waste is accidentally ingested by or entraps wildlife, it collects puddles of water in which malaria-bearing mosquitos breed, and it breaks down into tiny pieces called microplastics that are carried by air and water. We do not yet fully understand the risks the microplastics present in food chains for wildlife and for human health.[2]

Hakuna Matata; Hakuna Mfuko wa Plastiki

You might expect that the wealthiest and most environmentally progressive nations would be the first to address this plastics problem. Indeed, Denmark was the first country to adopt a bag tax. However, 31 sub-Saharan African nations have since begun to lead the world in plastic bag regulations.[3] With a GDP per capita about 15–20 times lower than that of the United States or Western European countries, Kenya is among the most aggressive, having banned most single-use plastic bags. The plastic litter problem in Kenya had reached a crisis point, literally covering the nation's roadsides and fields with litter and clogging its streams. Single-use plastic bags, along with bottles and all sorts of other plastic products designed to be thrown away are used widely in Kenya, as well as in low- and high-income nations across the globe. This bold step by the Kenyan government continues to meet with pushback from bag manufacturers and vendors, and it has provoked a predictable illegal black market in plastic bags. For the most part, however, it seems to be working, as evidenced by visibly less litter reported across much of the Kenyan landscape.[4] The Kenyan government has since adopted a policy that visitors may no longer carry plastic water bottles, cups, disposable plates, cutlery, or straws into the country's national parks and protected areas. Additional restrictions are being debated, with opposition by an oil industry group, the American Chemistry Council.[5]

"Design Is the First Signal of Human Intention"

As laudable as bans or taxes on single-use plastic bags may be in many instances, they do not really address the larger problem. Plastics are only one of many cases of accumulation of throwaway products. Producing single-use products is called a linear economy, which extracts raw materials (for example, petroleum), manufactures a product (plastic), distributes it to consumers for their use (bottles, bags), and then throws it away, usually after only a single use. If, like most plastics, the product is not completely biodegradable, it will accumulate somewhere in the environment, in a landfill, as litter on land, or floating in giant garbage patches in the ocean. It is estimated that only about 14% of plastic packaging is collected for recycling, and 4% of that is subsequently lost, while only 10% is effectively recycled. A similar percentage is incinerated. The remaining 72% is discarded, of which about 40% ends up in a landfill and 32% litters the landscape and seascape. About 8 million tons of synthetic polymer material enter the ocean each year. An estimated 150–200 million tons of garbage are now circulating in the Earth's oceans. By 2050 the mass of plastic polymers in the ocean is expected to grow so large that it will outweigh the mass of all of the fish in the world's oceans.[6] Something is wrong with this picture.

The measly 10% that is recycled may be used to make more bottles or perhaps plastic chairs or fleece sweaters. However, the current recycling economy depends upon the chair or sweater maker to figure out how to recycle products, or fibers of those products, which were designed for an entirely different purpose. Even with recycling, all of the plastic is eventually discarded, because the plastic polymers can only be recycled once or twice.

In contrast, a circular economy has reuse in mind when the original product is designed, produced, and distributed, so that most of it can more readily be recovered and reused in another production cycle (fig. 7.1). No system will be 100% efficient in terms of recycling of all materials, but losses from the cycle can be made small, because the emphasis on designing for reuse at the outset can raise rates of recovery, reuse, and recycling to a new level. Good circular economy design is needed in all sectors, including the vexing problem of one word: *plastics*.

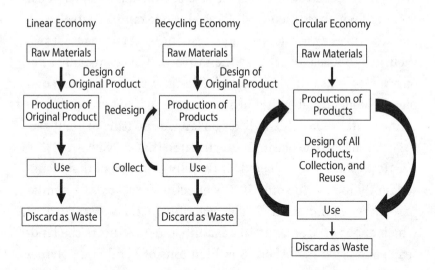

FIGURE 7.1 Differences among Linear, Recycling, and Circular Economies. Note that in the recycling economy only the original product is initially designed, whereas its potential recycling after its first use requires *redesigning* a means of reusing it. In contrast, the design of the original and subsequent uses of the product are part of the original design process in the circular economy, making recycling much more feasible, technologically and economically, and reducing the fraction that ends up discarded as waste. Drawn by the author

In addition to *Recovering* discarded plastic and engineering ways to *Reuse* and *Recycle* it, another essential component to solving the proliferation of plastic and other wastes is yet another R-word: *Reduce*. The less we produce and use in the first place, the easier it will be to recover, reuse, and recycle what we do use. Reducing plastics in packaging is a case in point. Amazon uses 100–500 million pounds of plastic per year for its packaging, most of which is difficult to recycle. While this accounts for less than 0.1% of global plastic production, it is still a big number for a single company. Packaging constitutes about 30% of the demand for plastic resins in the United States. The good news is that Amazon reports that it has phased out single-use plastic packaging at its fulfillment centers in India.[7] So, if it can be done there, why not globally? If a player like Amazon, with all of its internal R&D resources, could figure out how to do away with its single-use plastic packaging, the impact on the industry would reach far beyond its < 0.1% consumption of plastic. The alternatives for plastic must be carefully evaluated to make sure that their downsides are not worse than the plastic waste, such as alternative packaging that weighs significantly more, which would increase shipping costs and attendant greenhouse gas emissions. A convergence of knowledge from technology, social science, and economics is needed to design appropriate incentives—carrots or sticks or both—for engineers, architects, designers, retailers, and consumers to converge on solutions for creating this circular economy of plastics and other goods. It will also require markets and incentives for vendors and consumers to participate. An added bonus will be the reduced costs of proper garbage disposal and cleaning up messes from littering.

Architect William McDonough, whose quotation I have borrowed as the subtitle for this section,[8] espouses the philosophy that the designs that we create signal our intentions, not only in buildings designed by architects, but also in consumer products designed by engineers and marketers. If we intend for something to be reused or recycled, we should design it that way from the outset. Mimicking the cycling of water, carbon, nitrogen and other essential elements in nature, McDonough and his colleagues follow the flow of materials through products from the cradle to the grave and again to a new cradle.[9]

Just as the senescent leaves of a tree fall to the ground each autumn and release their nutrients to the soil, where they are then reabsorbed by tree roots and used for producing the next year's leaves (fig. 7.2), the fibers of a worn-out carpet, for example, can become the building blocks of new products that feed back into

FIGURE 7.2 Nature's Circular Economy as It Recycles Water (W) and the Nutrients Nitrogen (N) and Phosphorus (P). The light arrows show water cycling through clouds, rainfall, soil, plants, and back to the air as water vapor is released from leaves. The dark arrows show nitrogen and phosphorus cycling from plants to animals and from each to the soil, where they are again taken up by plant roots to build new leaves. Cycles in forests are on the left; cycles in agriculture are on the right. Diagram by the author

the market and the economy. This does not mean setting up carpet recycling bins for consumers to haul and dump piles of used carpets of many kinds. Rather, it means that the process of removing the old carpet and transporting it to a manufacturer who is already set up to use that specific type of fiber is built into the price of purchasing and installing a carpet. It also means that the carpet is designed in the first place with fibers that are meant to be reused and that can be readily separated from other components of the carpet. Carpet is a simple example of a product with an expected lifespan of several years before it needs to be replaced and the old one discarded or recycled, but the same concept of design for circularity can apply broadly to most products.

The notion of circular design is gaining momentum in the private and public sectors. The Save our Seas 2.0 Act passed by the US Congress in 2020 has an excellent definition of a circular economy that will hopefully guide policies in many sectors toward a more circular economy:

> The Save Our Seas 2.0 Act defines the term circular economy as an economy that uses a systems-focused approach and involves industrial processes and economic activities that:
> (A) are restorative or regenerative by design;
> (B) enable resources used in such processes and activities to maintain their highest values for as long as possible; and
> (C) aim for the elimination of waste through the superior design of materials, products, and systems (including business models).[10]

Previous chapters have already described key aspects of the needed circular economies. Our modern agricultural system has become largely linear, using fertilizers and other inputs to produce corn and soybeans in one place, transporting the animal feed to a distant place where livestock eat the grain, and then allowing much of the manure excreted by the animals to leak away, causing environmental pollution. A circular system would make sure that crop and animal production systems are integrated, geographically and logistically, so that the manure can easily and economically be returned to the cropland. That circularity would reduce pollution from release of manure into the environment and would help regenerate soil fertility and soil carbon. We should also collect food not eaten by consumers and use it to make compost, which is yet another way to recycle essential nutrients, rather than send it down the garbage disposal or into the landfill. That may sound a bit farfetched for Americans who have a throwaway culture, especially for stinky garbage, but I was pleasantly surprised to learn that South Korea successfully collects 95% of the food scraps from households and restaurants for composting, demonstrating that it can be done at a large scale in a modern, urbanized society.[11] A few US cities have nascent compost collection programs, too. Understanding South Korean and American cultures will be key to composting success, and green new deal thinking can contribute to that sort of convergence of cultural sensitivities with science and policy everywhere.

A Renewables Economy Must Also Be Circular

While discussing the deep decarbonization of our economy needed to keep climate change in check, we discussed the key role of advances in battery technology to store power from renewable sources of energy when the wind is not blowing and the sun is not shining. As a consequence, lithium and other metals needed for batteries and other parts of renewable technologies could become a potentially limiting resource of geopolitical importance. Key to the success of that vision will be recycling lithium and other elements from used batteries and electronics, which will require designing them so that these elements can be readily extracted from the used batteries.[12]

The recycling of essential elements needed by renewable energy technologies—including lithium, cobalt, and several metals

such as scandium, neodymium and dysprosium, that collectively are called *rare earths*—has potentially huge geopolitical
and national security implications.[13] The term *rare earth* is somewhat of a misnomer, because these elements occur widely but
usually at very low concentrations that make their extraction
technologically challenging and expensive. Mining those rare
earths can also cause a lot of pollution of soils, groundwater, and
streams.[14] Nevertheless, these metals are essential materials for
batteries, cell phones, military hardware, and many high-tech
uses. The United States currently depends upon imports from
China for about 80% of its rare earth supply. The US military is
investing in development of both mines and recycling technologies to secure rare earth metals for national security needs. It will
take many years, perhaps decades, to bring mines for lithium and
rare earths on line within the US to supply the needed minerals.
Even then, there may be insufficient capacity, and the mines will
create significant environmental concerns of their own.[15] One of
the reasons that China became the dominant global source of rare
earth metals was its willingness to accept the environmental devastation common at many of its mining sites. Developing large-
scale capacity to recycle lithium, cobalt, and rare earth metals
from scrapped electronics and to extract them from mining spoils
will also take time. But perhaps it will take less time than developing new mines and will create less environmental risk. Designed
recycling will be key to national self-reliance for the supply of
minerals essential to national security and a circular economy.

Green Capitalism and the Little Engine That Could

No, green capitalism is not an oxymoron. The circular economy
concept has many origins and permutations, but they share a vision, yet to be realized, of a market-based system that invests in
long-term solutions for the benefit of people as well as profit.

Therefore, a green new deal is needed that embraces and accelerates the concept of a circular economy, driven by entrepreneurialism and hard work and rewarded with the profit motive, but not laissez-faire, anything-goes, short-term greed.

A common thread of all of the examples considered here regarding circular economies—agriculture, food, plastics, carpets, batteries, renewable energy—is that most will require some sort of government intervention—carrots or sticks or both—to foster the transition to circularity. In most cases of dealing with waste, the cheapest and seemingly easiest short-term option for businesses and individuals is to discard and forget. As long as the monetary and non-monetary costs of trash and waste remain externalities of the marketplace, a linear, throw-away economy is likely to persist. Even with enlightened business leaders thinking about profitability and sustainability over the long term, short-term solutions for cheaply getting rid of wastes will remain very tempting. The costs of cleaning up pollution or enduring its negative human health, social justice, and environmental impacts, from plastic garbage to greenhouse gases, are then borne by taxpayers and citizens. Much of this pollution can be avoided by internalizing those costs into markets, which would help promote development of circular economies that avoid the production of waste in the first place, but that will usually require some degree of regulation or other financial incentives.

In a business context, sustainability emphasizes factors that value long-term over short-term profitability. This is why a green new deal is about more than just the environment and climate change: economics and social justice are also key aspects of long-term sustainability. Green capitalism posits that what is good for society, including environmental quality and social justice, will also be good for the bottom lines of companies that are in it for the long haul. In other words, environmental, economic, social, and long-term business concerns share a common need for sustainability.

A growing number of investment officers managing endowments of universities, private foundations, scientific societies, government pension funds and a growing number of mutual funds, exchange-traded funds (ETFs), and related investment options are embracing socially responsible investing (SRI) and environmental, social, and governance (ESG) criteria as part of their investment strategies. Only a few years ago this was a niche market for a small fraction of investors who cared about the social impacts of their investments. The small market and small number of investment choices meant that the investments were often less diversified and thus riskier than a broader portfolio of investments. When I started as an officer of the American Geophysical Union, our outside investment advisor mostly poohpoohed SRI and ESG options and warned us of greater risk and volatility. But this market has been transformed in the last five years and is becoming mainstream, with many options of high-quality and diversified ESG funds. Our current investment advisor recommends them wholeheartedly because they now have a history of performance that is as good as or better than traditional funds and large enough that the advisor can be selective when searching for quality.

BlackRock, the world's largest investment firm, managing some 9 trillion dollars of investments, claims that it is embracing ESG for all of its investments and that it will begin to exit certain investments, such as many coal companies, that "present a high sustainability-related risk."[16] Two examples followed that announcement: (1) BlackRock joined a shareholder's majority vote to require Procter & Gamble to commit to reducing its rate of forest cutting and attendant degradation of Canadian boreal forests for pulp that is used primarily for making soft toilet paper;[17] and (2) BlackRock joined other Exxon shareholders to elect three new board candidates who were opposed by management because they had pledged to steer the company away from fossil fuels and toward clean energy.[18]

Like the story of "the little engine that could," the Exxon share-holders' revolt was led by a tiny hedge fund called Engine No. 1 that thought it could (and thought it could, and thought it could . . .) get the big engines, like BlackRock, Vanguard, State Street, the New York State Common Retirement Fund, and the California Public Employees' Retirement System, to join its uphill battle. It did and together they won! To be sure, electing three environmentally minded members to the twelve-person Exxon board is a small victory, but it is a symbolic part of a larger trend of shareholder activism recognizing that fossil fuels may not be the best return on investment, both in the future and even now.

In addition to the E of ESG, corporations are also being ranked by investment advisors according to the workplace social (S) culture and governance (G). A growing number of corporations see value in providing their employees decent wages and benefits and advancing women and people of color to senior positions. Companies with satisfied workers, a diverse workforce and leadership with ideas and perspectives, transparent governance structures, and long-term sustainability plans tend to be more profitable in the long term, which is attractive to investors and which, in turn, makes capital more available to those companies.

In the next chapter, I discuss the importance of diversity in academia, such as the more diverse questions and approaches to science that a diverse faculty and student body bring and which make our science more useful and productive. I know academia best, but the same concepts apply to the business world. The ESG movement is putting pressure on corporations to diversify their boards and their senior executives, so that those who have historically been left out become part of decision-making processes. This diversification is an essential element of the transformation from a capitalist system designed to advance the self-interests of the predominantly white male privilege holders to one that pursues corporate policies for promoting the well-being of communities that historically have been without voices in board rooms and C suites.

While encouraging, the ESG market is far from perfect. Data are often incomplete, the benefits of ESG criteria are not always clear, and one must always be on the lookout for greenwashing (that is, false or questionable claims meant to make a company's environmental record look better than it really is). The interest in ESG investing would be more productive if more transparency in corporate reporting of ESG criteria and performance were required. Beyond ESG, more proactive investors are embracing the concept of impact investing, which creates and directs investments specifically targeted toward good business opportunities for advancing renewable energy, increased efficiencies, improved water management technologies, regenerative agriculture, and other contributions to the circular economy vision.

These and other examples of responsible corporate actions suggest a movement toward mainstreaming environmental and social concerns, albeit with imperfect track records and at a pace slower than needed. In chapter 4, for example, we observed that Walmart could be lauded for requiring nutrient management plans on farms that supply some of its grocery store products, for streamlining its trucking system for efficiency, and for installing solar panels on many of its stores, while also noting that it still falls short with regard to employee benefits and sales of single-use plastics and packaging. Walmart is also one of 115 major purchasing organizations around the world, collectively representing $3.3 trillion in procurement spending, that have joined a nonprofit organization called the Carbon Disclosure Project (CDP). The CDP runs a global disclosure system for investors, companies, cities, and states to report on their environmental impacts, such as climate change, water security, and deforestation. The CDP reports that more than three hundred corporate suppliers recently provided disclosures related to the use of commodities that are associated with deforestation, but of those, only 17%

have set any sort of target related to reducing their deforestation impacts.[19] Although an improvement over previous years, this progress is too slow. On the other hand, you cannot change what you do not acknowledge and measure, and this process of public disclosure of measured impacts is an encouraging sign, with trends going in the right direction. Indeed, additional momentum for disclosures regarding climate change is coming from a growing number of asset managers[20] and from BlackRock. CEO Laurence Fink is now calling on the companies in which Black-Rock invests "to disclose a plan for how their business model will be compatible with a net-zero economy. . . . We expect you to disclose how this plan is incorporated into your long-term strategy and reviewed by your board of directors."[21] Perhaps the three new Exxon board members whom Fink helped elect will repeat his message frequently to their fellow board members.

Greenwashing: "Hypocrisy Is the First Step to Real Change"

Despite several positive signs of mainstreaming environmental concerns by corporations and investment firms that wield enormous influence due to their large market shares, we should not be so naïve as to think that these corporations are doing all that they could to reduce greenhouse gas emissions and other environmental impacts, or to promote social justice in their work environments. There is much more to do. For example, a group of CEOs belonging to a group called the Business Roundtable signed a statement pledging to invest in their employees, protect the environment, and deal fairly and ethically with their suppliers.[22] However, as Marc Benioff, founder of Salesforce and one of the statement's signatories, said in response to criticism about living up to those pledges, "I've seen from my own viewpoint a systemic change in how C.E.O.s behave over the last 20 years. I never said it's a revolution, but I said it's an improvement."[23] *Revolution* is a loaded word meaning different things to different people, but we certainly need more than the slow incrementalism that Benioff describes. Moreover, there are still plenty of individuals and corporations that have not signed onto the Business Roundtable pledge or to the Carbon Disclosure Project and continue to ignore environmental and social concerns. The private sector has to be part of the solution, and so it is highly significant that there are processes and some momentum for private sector responsibility upon which meaningful partnerships with green new deal thinking can build further and faster.

At the same time, we must remain on the lookout for corporate hypocrisy. In response to a question about guarding against corporate greenwashing, Hunter Lovins, energy specialist, president, and founder of the nonprofit Natural Capitalism Solutions replied:

> Yes, but I think greenwashing is good. Hypocrisy is the first step to real change. . . . If a company makes a claim about something, then you can hold them accountable. . . . And then as they make steps to bring their performance in line with what they're marketing, they actually see the benefit of that improved performance, and it becomes something they integrate into their business for real.[24]

Now that they have made public pledges, those Carbon Disclosure Project CEOs and Business Roundtable CEOs can be held accountable to living up to them, because we need more than the incrementalism that Mr. Benioff calls "an improvement."

Bounding Unbound Capitalism

I would be remiss if I didn't acknowledge that many green new deal thinkers are far more circumspect about green capitalism. Naomi Klein, for example, disparages the "tepid, market-based solutions" offered by "neoliberal centrism." I mostly agree with her assessment that the drive for endless growth and profits stands squarely opposed to the imperative for a rapid transition off fossil fuels. It is absolutely true that the global unleashing of the unbound form of capitalism known as neoliberalism in the 80s and 90s has been the single greatest contributor to a disastrous global emission spike in recent decades, and the single greatest obstacle to science-based climate action since governments began meeting to talk (and talk and talk) about lowering emissions.[25]

Indeed, I would add that the "unbound form of capitalism" that Klein blames for lack of adequate action on reducing greenhouse gas emissions is also largely to blame for the large increase in wealth disparity since the 1980s shown by economist Thomas Piketty (see fig. 2.2). Klein points to democratic, quasi-socialist countries "like Denmark, Sweden, and Uruguay" as a model for "the most visionary environmental policies in the world." I agree that the United States and other countries could learn much from those examples, but I would also note that they fall along a spectrum of mixed economies, with varying degrees of free-market capitalism, government regulation, and government-provided social safety nets. Capitalism itself is not the culprit, as profit motive still drives private sector businesses in countries like Sweden. Where along that spectrum the US and other countries should fall involves cultural as well as economic and environmental determinants. The convergence of social science with other green new deal stakeholders will help us find that sweet spot for each culture and country now and in the future as circumstances and cultures change.

In my opinion, it is the "unbound," largely unregulated form of capitalism that Klein calls the neoliberalism of the 1980s that must be reined in. Appropriate guardrails can provide protections against perpetuation and expansion of self-interests while still harnessing the potentially powerful positive aspects of markets to help achieve green new deal objectives. This is not a "tepid" approach, but rather is an essential strategy to have as many forces as possible, including targeted market forces, working toward effective change. This is not to say that I put blind trust in business leaders like Larry Fink to get the job done. They will need encouragement, guardrails, and perhaps an occasional kick in the pants from appropriate government policies. Nevertheless, I see business leaders as important key players with much expertise to contribute to the development of functioning low-carbon-emitting, circular economies, provided their actions match their words, and do so quickly.

Molding market incentives to be consistent with long-term sustainability goals, such as net zero-carbon emissions, circular economy, and increased diversity at all levels of business, is a far-reaching goal of convergent and transdisciplinary approaches to a green new deal. A private sector with more long-term and diverse perspectives will be a necessary partner with the natural science, social science, economics, engineering, and community engagement convergence needed to advance successful technologies and practices of a circular economy. Healthy skepticism is in order, calling out hypocritical greenwashers and converting them to genuine green new deal partners, and transforming unbound capitalism to green capitalism. The bold visions of a green new deal require major transformations that will need allies from both public and private sectors—after all, it is a really *big* deal.

Recommendations

The Green New Deal (H. Res. 109, 116th Cong., 1st Sess. [introduced February 7, 2019]) acknowledges the role of business and industry as partners in building wealth while addressing social environmental concerns such as climate change:

> *directing investments to spur economic development, deepen and diversify industry and business in local and regional economies, and*

build wealth and community ownership, while prioritizing high-quality job creation and economic, social, and environmental benefits in frontline and vulnerable communities, and deindustrialized communities, that may otherwise struggle with the transition away from greenhouse gas intensive industries.

However, the GND resolution does not delve into topics like recycling, the circular economy, long-term versus short-term profitability, or socially responsible investing. This is an oversight, but one that we can rectify here with a few high-level, green new deal type recommendations:

Innovations are needed that reduce the production and use of virgin plastics and that design plastic products or their alternatives for recovering, reusing, and recycling in a circular economy. Plastic products of various kinds will still be produced because many have extremely useful properties that would be very difficult to replace. However, some novel alternatives for many uses of plastic will reduce reliance on plastics, and circular designs will ensure effective recycling of the products that continue to be made of plastic. As is the case for deep decarbonization, major changes needed to reverse current unsustainable proliferation of plastic wastes will likely require a combination of approaches, including incentives, regulations, and R&D in both public and private sectors. Banning single-use plastic bags may be commendable in some instances, and bolstering efforts to recycle plastic bottles and the plethora of other current plastic products should be encouraged with appropriate policies. However, those efforts will be inadequate to reverse the growing global accumulation of plastic waste. A redesign of plastic production and use consistent with a circular economy will be necessary.

Additional funding and tax incentives are needed for public and private sector R&D in a wide range of circular economy technologies, including social and economic impediments and opportuni-

ties for adoption. In particular, a circular design will be necessary for sustainable use of lithium, cobalt, and several rare earth elements needed for renewable energy technologies and energy storage in batteries.

Where circular technologies already exist, subsidies and other incentives for early adoption may be warranted. For example, just as tax deductions and rebates are offered for purchases of energy-efficient appliances and vehicles, they could be offered for products like carpets and electronics that meet circular economy design criteria.

More transparency and improved reporting on the environmental, social, and governance (ESG) policies and track records of publicly traded corporations are needed and should be required by the Securities and Exchange Commission and other agencies with relevant regulatory authorities.

Innovations are needed to incentivize corporations to disclose and mitigate the environmental and social impacts of their supply chains of commodities. Such initiatives may include better informing customers of the consequences of their consumer choices and may include incentivizing corporations through outreach, regulations, and tax incentives.

8

Whither the Academy? A Horse of a Different Color?

Harnessing science, as discussed in chapter 1, suggests a horse whose power is directed toward a pathway of progress. That's a pretty good metaphor for science's contribution to society, but it's time to consider that we also need a horse of a different color. The way we do science needs to change to support a green new deal. And let's not forget the diversity of the jockeys.

Investment in Science Always Pays, but...

Researchers devoted to top-notch, basic research in their disciplines are expanding human knowledge, which yields countless benefits for humanity. Although their topics may seem esoteric and their jargon often impenetrable, that basic research frequently ends up forming the basis for unanticipated useful applications many years or decades later. Albert Einstein's theories of general and special relativity are difficult to grasp, to say the least, and they might seem to have little practical effect on our daily lives. However, the theory of relativity is essential for calculating the positions of satellites relative to the Earth, which is the basis for the GPS systems that we now take for granted as a very convenient and practical feature of our smart phones, not to mention their crucial applications for natural resource management and for the military.

The smart phones themselves were enabled in part by the miniaturization of electronics needed to take astronauts to the moon and back. Ironically, today's smart phones are more powerful than the computers aboard the 1969 Apollo spacecraft, but the cell phone would likely not have developed nearly as quickly without the moon shot. The benefits far exceed the convenience of being able to keep in touch with the babysitter. When I lived with subsistence African farmers in the days before cell phones, they were at the mercy of merchants who arrived in the village with a truck, offering to buy the farmers' corn and cotton for a nonnegotiable price. The farmers had no way of knowing the current market price outside of their villages (there were no landline phones or newspapers either), or whether another merchant might arrive to offer a better price, so they often sold to the first merchant at whatever price he offered. Now, someone in the village has a cell phone, so the sons and daughters and the grandchildren of the farmers that I knew can call to find the best price for their produce and who is coming to buy it. They also use phone apps for agronomic and health care advice. The scientists and engineers who were miniaturizing electronics for NASA could never have imagined the future practical value of their inventions, which are making many peoples' lives better.

Similarly, scientists with government grants for basic research in the early 1960s first discovered messenger RNA and its role of translating DNA within cells to produce enzymes that carry out life's functions. I bet they never dreamed that the basic knowledge they had produced about the machinery of how cells work would lead to the development of an effective COVID-19 vaccine in 2020, saving millions of lives and preventing debilitating "long

COVID" symptoms in millions more. Modern big pharmaceutical companies deployed their resources and their practical, applied research knowledge to translate the basic science into effective and safe vaccines, thus also playing a huge role in their record-breaking rapid development. There is more work to do, as the scale of vaccine production needs to be expanded to make them available to the developing world, including production sites within those countries. Combinations of public and private investments in both basic and applied science pay unexpected dividends and always will.

That model of investing in basic science performed by highly specialized experts at universities and research institutes, and then waiting for the payoffs to emerge at some later date as engineers and entrepreneurs find ways to apply the science, will continue, as it should. Many scientists work best within the comfort zones of their specialized disciplinary research. Yet that model alone is no longer sufficient. Disciplinary science will continue to be valued for its contribution to humanity, *but* the challenges of climate change and other forms of environmental pollution, the sustainability of economic productivity, and social justice issues demand a more urgent and direct role for interdisciplinary science. A growing number of natural and social scientists are collaborating with each other and partnering with communities to seek solutions in interdisciplinary and transdisciplinary teams. Similarly, a growing number of scientists, especially many younger ones, explain they were drawn to science not only for the wonderment of discovery, but also to help make a difference in the world now, as society grapples with vexing problems. Convergent science, as described in the preface and in chapter 1, offers an exciting additional career path option for scientists. For many, its appeal lies in the opportunity to learn beyond a single discipline, to engage with experts from other disciplines, and to work with nonscientists in their communities to attack those vexing problems together.

Unfortunately, this new trend in interdisciplinary and community-integrated science is often not rewarded or encouraged in academia. The structures of academia and the broader scientific enterprise are based on an old model that, in its attempt to protect the purity of science, often fails to reward scientists for venturing beyond their narrow disciplinary specialties or for engaging with society. In fact, such efforts are often specifically discouraged by traditional disciplinary criteria for awards, promotions, and tenure.[1]

When science is isolated from the broader society, it also retains structures that offer preferred access to those groups that have historically enjoyed the privileges of higher education. Hence, the issue of tradition affects both the interdisciplinarity of research and the diversity of researchers. A perpetuation of lack of diversity among scientists also limits the diversity in the way science is done, the questions that scientists ask, and the solutions that they seek. Therefore, we need a systemic change in academia that will enable new partnerships of diverse scientists and community members to flourish in the coproduction of new knowledge needed to create a sustainable future.

Many who seek improvements in diversity and inclusion in academia are substituting the terms *minority communities* or *underrepresented communities* with the more informative term *minoritized communities*. This expression acknowledges that the status of these groups has less to do with their numbers per se (that is, that they are a minority or are underrepresented relative to the population at large) and more to do with the consequences of historical withholding of their rights, opportunities, and privileges by the majority.[2]

When scientists engage with communities, more and more people begin to learn that scientists do much more than work in lab coats on esoteric research (as important and meaningful as that research may be). Working with members of the community to measure their air, water, and soil quality, to assess their vulnerability to floods and heat waves, and to brainstorm about how to solve these problems, not only benefits the scientists and the community members, but also exposes young people in those communities to a new vision of what scientists do, how they look, and how science could be relevant to their lives. When the public, including children, parents, and community leaders, see the value of engaging with diverse scientists in equitable exchanges of experience and knowledge, they will start seeing valuable and rewarding pathways in science, technology, engineering, and mathematics (STEM) for their young people.[3] We all have an idea of what doctors and lawyers do, but how many children, or their parents for that matter, have a clue about the daily work of environmental scientists and geoscientists?

I had the honor of introducing and interviewing Lisa Jackson as the 2018 Presidential Lecturer for the American Geophysical Union (AGU). She was the first African American administrator of the US Environmental Protection Agency and then became the vice president for Environment, Policy and Social Initiatives at Apple. She recounted a personal story of her early career choices. Despite encouragement from her family to go to medical school, she informed them that she wanted to become an environmental engineer. "That's nice honey," her grandmother responded, a bit befuddled, "but why do you want to drive a train?" I could imagine my grandparents saying the same thing. Fortunately, she was able to convince her family to support her unusual journey into a field that they initially misunderstood, not realizing that it existed as a viable career path. The rest is history, as they say.

Prospective students also may not know what opportunities are affordable. A Hispanic professor and friend described to me his conversation with an undergraduate student of color who was presenting a poster at an AGU meeting about her undergraduate research project. After discussing the poster, which impressed him favorably, he asked whether she had plans to apply to a graduate school, where she could continue to pursue interests similar to those that inspired her undergraduate project. No, she replied, because she couldn't afford it. He then explained to her that most master's degree and doctoral programs involving research in the environmental sciences and geosciences are funded by research grants that cover tuition and pay the student a modest but livable stipend. That was news to her. In the end, she applied to his university and was accepted with a stipend and is now doing research for a degree. I suspect that having a professor who is himself a scholar of color will improve the chances that she will find a welcoming and inclusive graduate school experience, although that is not a guarantee. Nor should such inclusiveness and outreach to students from diverse backgrounds be limited to a few professors of color who have managed to overcome barriers.

If we want more children from historically minoritized communities in STEM, they have to see that there are more opportunities available to them than being doctors, lawyers, or even engineers, and that those opportunities can make a difference for their communities. One African American student in a course I co-taught on sustainability science took on a class project to merge geographic information system (GIS) data on land elevations along the coast of the Eastern Shore of Maryland with demographic data on population by race and income. He used those tools to show that a large fraction of the low-lying areas most vulnerable to storm surges and sea level rise was inhabited by African Americans with low incomes. He was motivated by science that he saw as relevant to his communities, and fortunately for him, he found his way into a master's program in geoscience in which he is learning to apply those skills. Because of who he was, he pursued a science question that no other student had asked. Sadly, he is the exception rather than the rule, as the geosciences in the United States have an abysmal track record for conferring degrees to students who identify as minorities (fig. 8.1). There are many complex reasons for this, but one that clearly stands out is

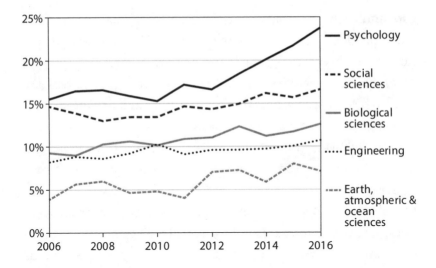

FIGURE 8.1 Doctoral Degrees Conferred to Underrepresented Minorities in the United States. The percentage of doctoral degree recipients who identified as Hispanic or Latino, American Indian or Alaska Native, or Black or African American is shown for each discipline. Drawn by the author from data reported by National Center for Science and Engineering Statistics, "Women, Minorities, and Persons with Disabilities in Science and Engineering," https://ncses.nsf.gov/pubs/nsf19304/data

that many young people of color have not yet been exposed to role models in the geosciences that resonate with them. Getting more scientists partnering on-site with communities to address their environmental and social concerns will enable a new generation of children and their parents to see scientists as potential role models, especially if some of those role models look like them.

The problem goes beyond initially recruiting minoritized students to STEM fields. Once interested, students and early career scientists often are not offered the support and opportunities needed to flourish. They often lack inclusive learning and working environments in which they feel valued and enabled to succeed. The Meyerhoff Scholars Program at the University of Maryland Baltimore County (UMBC) is one of many innovations of its retiring president, Freeman Hrabowski III.[4] That program fosters a highly supportive, tightly knit learning community of undergraduate students from all backgrounds who aspire to pursue doctoral degrees in the sciences or engineering and who are committed to increasing the representation of minoritized communities. It provides resources for summer bridge programs, internships, and conferences. According to the National Science Foundation, UMBC graduates more Black students who go on to earn doctorates in the natural sciences and engineering than any other US college. One of Hrabowski's legacies will be his demonstration that this kind of success is possible.

These success stories are great, but unfortunately there are still many impediments in academia and other research institutions that discourage engagement of scientists from all backgrounds in solution-oriented, community-based science. Transdisciplinary and community-based science requires the building of trust among the collaborators and learning to speak each other's language. That trust building takes time and requires mutual understanding of what brings each participant to the collaboration. Community leaders seek some form of positive change for their communities. The scientists want that, too, but they must also demonstrate their scholarship to the university, based on the university's definition of what constitutes scholarship. Promotions and tenure usually depend largely on the number of papers published and the number of research dollars awarded by granting agencies within a certain number of years. Administrators and other faculty in the same academic department, especially the more senior ones who came up through the ranks following the traditional disciplinary model, may not see the value in research projects that are partly outside of their discipline, that are more solutions-oriented than the traditional view of scholarship based on basic research, or that require more patience to see results.[5] Those views are gradually changing, but it is a cultural change currently proceeding at an insufficient generational pace.

Agents of change, however, like the new generation of young scientists whose stories I tell, with support of visionary scientific societies and institutional leaders, are accelerating the pace of change. My university has adopted criteria for promotion and tenure that acknowledge the value of interdisciplinary integration and application of science. Putting those criteria into practice has met some resistance by some faculty accustomed to the old ways, but it is also being supported by commitments of the senior leadership, pushed by the junior faculty, and cheered on by the graduate students and postdocs. A growing number of workshops and symposia at annual meetings of scientific societies are questioning traditions of promotion and tenure. My examples come from the American Geophysical Union (AGU), which I know best. The AGU has also devoted significant funding to its Thriving Earth Exchange program, which has carefully vetted over 100 community groups and matched their expressed needs with scientists who are looking for collaborative, community-based science opportunities.[6]

Funding agencies like the National Science Foundation (NSF) are supporting projects that actually require community participation, like the aptly named "Coastlines and People" program. It seeks to "support potentially ground-breaking investigations on Coastlines and People that cross disciplines, involve stakeholders and local communities, and integrate broadening participation into the values of all activities and research." The Department of Energy has proposed integrated field laboratories that would gather climate data in cities and build bridges to urban communities, including collaborations with minority-serving universities, such as Historically Black Colleges and Universities. Likewise, NSF's "Navigating the New Arctic" initiative encourages scientists to enlist Indigenous communities in the "co-production of knowledge" by engaging them from project planning stages onward. Despite these good intentions, it has been a rough start, because many scientists do not yet have experience, skills, and comfort with the concept of co-production of knowledge with nonscientists. An NSF program officer lamented, "We made a mistake in assuming the scientists knew what that meant . . . we definitely have a lot of work to do to make sure that Arctic sciences is diversified and equitable."[7] As long as the necessary leadership and commitment continues, I am optimistic that we can overcome these growing pains and will see community-based science flourish with very positive outcomes.

Who Holds the Reins?

A still more profound and urgent change will be essential for harnessing science to address interrelated problems of the environment, economics, and justice. After all, the universities that host our centers of scientific excellence were founded on a colonial, hierarchical model of privilege that embodied the sexism and racism of their day. This tradition continues to discourage, sometimes blatantly and sometimes insidiously, entry by people who do not look like the academic hierarchy.

The modern universities of the West can be traced back a millennium. Some were essentially guilds for learning specific crafts from masters of those crafts. Many were affiliated with the Catholic Church and later the Lutheran Church. They have been resilient through great turmoil and have stood as important symbols, to varying degrees, of academic freedom and open inquiry. However, they were also designed to perpetuate the privileged classes that created them and that had access to them. Even in countries where universities are now state-run and tuition-free to all of their citizens, centuries of academic traditions often determine which types of students and faculty are advanced to what opportunities and how. Merit plays a role, as it should, but we would be fooling ourselves to believe that the reins of academia are held entirely or even mostly based on a system of meritocracy.

Women have finally made significant inroads into this mostly white male bastion, but it has been a long and difficult journey, and it is still far from over.[8] Historically minoritized groups are only at the very initial steps of that journey. The problem is not so much that those of us in charge are personally discriminatory, although we all have unconscious biases and most of us took good advantage of several privileged opportunities during our careers that we may or may not recognize or acknowledge. To their credit, many concerned scientists and educators are making well-intentioned incremental changes, such as forming justice, equity, diversity, and inclusion (JEDI) committees, offering implicit-bias training to search committees, and investing in more aggressive recruiting. While laudable and significant, these incremental steps have yet to move the needle for people of color in the geosciences and several other science fields, because they do not address the fundamental structural problems of academia that have sexist and racist implications.

Learning from efforts to address structural discrimination against women may shed some light on what is needed for similar structural changes to address racial discrimination. As an elected officer of the American Geophysical Union, I participated in evaluation of some difficult ethics violation cases, including evidence of women, usually young women, who dared to come forward to accuse a more senior male colleague of sexual harassment, usually the professor who advised her and held the keys to her professional future. Graduate students working on master's and doctoral degrees can change advisors, but it often is not easy, it is discouraging to say the least, and the disruption can cost them years of effort and loss of funding. The hierarchical academic system empowers professors, especially the more senior professors, who not only are the gatekeepers for conferring degrees, but who can also greatly influence the employment options of students following graduation. They can also sway the tenure and promotion decisions for junior faculty. Sexual favors are sometimes the price of getting such endorsements.

While one might claim that the problem is just a few bad apples behaving badly, the culture of academia has been to avoid sticking one's nose into another professor's affairs, and so those bad actors too often enjoy a complicity of silence among their colleagues. Many scientific societies and some universities are now offering witness training, to help those who observe harassment, bullying, and discrimination to know how to act. One senior faculty member of my university expressed astonishment to me after completing the training; he said that he had never thought that it was his business to get involved if he saw something. My response was to admonish him to remember the slogan of the US Transportation Security Administration looking for danger at airports: "If you see something, say something!"

Another problem is that I am convinced that many of us are clueless about what is going on under our noses. Some of my women colleagues don't buy that explanation, and I respect their opinions, but I maintain that the obliviousness of absent-minded male (or female) professors should not be underestimated. Being out of touch is not a good excuse by any means, but it happens nevertheless. Actually, it isn't limited to old professors. Once I became more aware of the issues through my role as a scientific society officer, I dug up old memories from my distant past, and realized for the first time how clueless and naïve I had been as a student.

As an 18-year-old boy entering a freshman class in a small liberal-arts college in the mid-1970s, I was privileged to have my eyes opened to many new ways of viewing the world. Feminism was a surging movement on such campuses then, and I engaged in many discussions in and out of the classroom about discrimination against women in education, the workplace, and in society in general. However, I have no memory of discussing what we now call sexual harassment. My guess is that the women were discussing such matters among themselves, but it wasn't discussed publicly as a general topic and certainly not with respect to specific local cases. So when I learned that some of my classmates were sleeping with their professors, I jumped to the conclusion that it was consensual and that they either were truly attracted to those men or were purposefully currying favor to improve their grades. It never occurred to me at that time that when the professor presses for sexual favors, the student may be in an almost impossible situation. If the course was a requirement for the major that she was pursuing, or if it was too late in the semester to drop the course without getting an F, her options were limited. There were no persons or offices set up in those days to report the professors' behavior, and they enjoyed deniability of witnessless abuse. I am still perplexed and ashamed why I never thought of that possibility back then. I guess I was just clueless, and I was not alone. I present this explanation not as an excuse, because it is not, but to demonstrate that even the most clueless among us are able to learn, eventually, that such behavior cannot be tolerated.

In my senior year, I finally learned to appreciate how vulnerable some of my classmates were. A classmate confided in a few of her friends that a senior administration official of the college was pressing her to have an affair with him. She had agreed to babysit his kids to earn a little extra spending money, and he hit on her when driving her home one evening. She refused, but he continued to contact her and to stalk her for months thereafter. Upon hearing this, none of the small group of friends with whom she had confided thought for a moment that she should report his behavior. This was before the Title IX laws, before the term "sexual harassment" was even in our lexicon, and the college had no mechanism that we knew about for dealing with such cases. We all felt instinctively that if she reported his behavior, it would be his word against hers, and the power that he wielded would ensure that she would be the one told to leave campus, not him. We did our best to console her and to offer ourselves as chaperones when she needed to walk around campus, hoping that he would eventually lose interest if he could never find her alone. Fortunately, she persevered without giving in to his advances, and she graduated that spring, leaving campus with her degree in hand, but no doubt with scars of the injustice and humiliation that she was forced to endure. While bad enough, this anecdotal personal story could have had a worse ending; indeed, many of my women colleagues can tell horrific stories, with endings ranging from enduring and mostly moving on, as my friend did, to more tragic consequences that changed lives and careers.

Most US colleges and universities now have human resources departments with clear procedures for adjudicating harassment complaints. This option, which was not available to us in the 1970s, is helping to gradually increase the percentage of women coming forward with their cases, but there are still many cultural impediments to coming forward within academia and society at large.[9] One of several key factors is for institutions, within and beyond academia, to demonstrate that there will, indeed, be consequences for the perpetrators. Until recently, that was rarely the case, and it still does not happen consistently.

Moving the Needle: Redefining Scientific Misconduct

My proudest moment while serving as president of AGU, was to announce in 2017 to our approximately 60,000 worldwide members that we had adopted a new ethics policy, which defined harassment, bullying, and discrimination as scientific misconduct.[10] Until then, scientific misconduct was limited to such unethical behavior as plagiarism or falsification of data. Harassment was considered social behavior that lay outside of the scientific process. And yet if harassment, bullying, and discrimination discourage those who are trying to become scientists, then that affects the scientific process by effectively excluding segments of our society from the privileged halls of academia and research institutes.

That is misconduct, not only because being discriminatory and exclusionary is morally wrong, but also because it reduces the power of science to be harnessed to address society's vexing problems. How does that conclusion follow? The questions that scientists ask and the way they go about seeking answers to those questions depends upon who the scientists are. Growing up in Montana, I was most interested as a young scientist in studying natural ecosystems with limited human influence, while several of my colleagues who grew up in big cities were attracted to studying the ecology of urban environments. Likewise, our interest and abilities to relate our scientific expertise to the interests and needs of local community groups may depend at least in part on our gender, race, ethnicity, sexual orientation, and economic class. The African American student in my sustainability science class was motivated to ask whether the impacts of sea level rise and storm surges are borne disproportionally by people of color in our state. A more diverse science workforce will ask more diverse questions and use more diverse and innovative methods and partnerships to answer those questions for the benefit of society. Even in a disciplinary field as remote from community affairs as, say, particle physics, the means and outcomes for training young physicists will vary depending on who is doing the training.

The visionary redefinition of scientific misconduct by AGU has had a snowball effect, as other scientific societies, government agencies, research institutes, universities, and field research camps have adopted all or part of our ethics policy. Our sister society in Japan, where participation of women in science is remarkably low, took notice during joint society workshops with AGU and has reevaluated its ethics policies. There remains resistance, to be sure, but it is gratifying to see the needle move due to the concerted actions of many.

While I had the honor of announcing the new ethics policy, it was the culmination of efforts of scores of AGU members who had organized town hall meetings, workshops, and conferences to raise awareness on the topic. The AGU president before me established and charged a task force with studying and proposing updates to our ethics policy. The proposed policy was thoroughly vetted at several levels of governance in the organization. By the time we got to my announcement, we had broad buy-in from our members. Coincidentally, my announcement came only months after the Harvey Weinstein case emerged in the popular press. Thanks to the work of many dedicated volunteer members over several preceding years, we were ahead of the curve of the #MeToo movement with an action that introduced a significant structural change to academia. I wish I could say the same for improved representation of historically minoritized communities in science.

Investment in Science Pays; Investment in Inclusive Science Pays More

Following the model used to revise our ethics policy, AGU created task forces on justice, equity, diversity and inclusion (JEDI) and on talent pool development. We are now implementing their recommendations, systematically adding JEDI considerations to everything we do, including governance, publications, meetings, honors, student mentoring, and career development. But unlike our proud moment of being ahead of the #MeToo movement curve in 2017, I cannot honestly say that we were ahead of the curve when the brutal 2020 killings of George Floyd, Breonna Taylor, and Ahmaud Arbery refocused our attention on the history of systemic racism, which continues to have persistent, palpable influences on who has privileged access to academia and the broader scientific enterprise today.

With this 2020 vision, and our growing awareness of the chronic brutality and institutionalized subordination suffered by communities of color for far too long, the urgency of going beyond incrementalism is all too apparent. Making the scientific enterprise more inclusive will require a structural change that redefines the way we do science. The convergent science discussed throughout this book is one such change that will require enabling and rewarding more place-based, community-engaging, solutions-seeking projects in which the scientists are part of an equitable co-learning and co-knowledge generating team.[11] It is a way of getting more kids of color interested in STEM by seeing scientist role models in their communities partnering to solve problems important to them. It is also a means of doing science in a manner more inclusive and supportive for those early-career scientists who have managed to make it through academia's gates. We must go beyond the NSF definition of convergence of "expertise from different disciplines" to include expertise, approaches, and perspectives from different communities. As science becomes more inclusive, the questions we ask and the solutions we find will become more relevant to the societal needs that we serve, which will further reinforce inclusivity, interest, and trust in science in a positive feedback loop.

Most importantly, the leadership of institutions must lead. Most of those leaders enjoyed considerable privilege to get where they are today, but as Professor Vernon Morris explains, what matters is how we use those positions of privilege going forward:

> We have to accept the presence of privilege and that it will persist. We do not have to accept the use of privilege to: i) exacerbate inequity and buttress systems of oppression, ii) maintain practices that maintain an imbalance of access or inclusiveness, or iii) silence

the marginalized. Instead, privilege can be used to challenge the imbalances of low expectations and deficit model thinking, create opportunities for inclusion, and reduce the adherence to exceptionalism as a defensive posture in discussions of structural racism. This is where academic leadership can make the greatest impact.[12]

While some change is already underway at funding agencies like NSF, scientific societies like AGU, and universities across the world, much more could be done to raise to a new level the efforts to change the way science is done and to break down the academic structures that have systematically discouraged people of color and women. A green new deal would address disparities in access to education, starting from pre-school all the way through graduate school and beyond (pre-K to gray). At the level of government-supported universities, it would accelerate the changes already underway to reward interdisciplinary, community-based, solutions-oriented science in tenure and promotion policies. We must redouble our efforts and resources to increase and retain diversity in science because, while investment in science pays, investment in inclusive science pays even more.

Recommendations

The Green New Deal (H. Res. 109, 116th Congress, 1st Session [introduced February 7, 2019]) recognizes the need for both higher education and community involvement, and, if you read carefully, it also connects those dots:

> providing resources, training, and high-quality education, including higher education, to all people of the United States, with a focus on frontline and vulnerable communities, so that all people of the United States may be full and equal participants in the Green New Deal mobilization; . . .

building resiliency against climate change-related disasters, such as extreme weather, including by leveraging funding and providing investments for community-defined projects and strategies; . . .

mitigating and managing the long-term adverse health, economic, and other effects of pollution and climate change, including by providing funding for community-defined projects and strategies; . . .

ensuring the use of democratic and participatory processes that are inclusive of and led by frontline and vulnerable communities and workers to plan, implement, and administer the Green New Deal mobilization at the local level.

We can be even more explicit about connecting those dots in the following recommendations:

Universities must rethink their criteria for tenure and promotion, giving full credit for transdisciplinary research and community engagement as scholarly pursuits worthy of career advancement.[13] Many universities outside the United States are controlled by national governments, many are controlled by state governments as in the US, and many are private, so there is no single authority for adopting these recommendations. Moreover, these changes will require a cultural shift within universities and research institutes. Government policies can influence the culture of doing science, but the needed cultural shift will also require tireless champions of change at leadership positions as well as champions at every career stage.

Government agencies that fund research, such as the National Science Foundation and the National Institutes of Health in the US should expand their funding of convergence research. Beyond more funding, however, the agencies should expand assistance to the research community to learn *how* to successfully engage in community-based science.

Beyond establishing committees on justice, equity, diversity, and inclusion (JEDI), institutions of all kinds, and especially those in edu-

cation and research, need to take concrete steps to recognize and break down vestiges of structural racism and sexism from top to bottom in their governance and management structures. This includes bystander training, clear statements and enforcement of ethics policies, enhanced recruitment and inclusive retention initiatives, and mainstreaming JEDI considerations in all institutional functions. It also includes equitable partnerships among predominantly white first-class research institutions (also known as R1 institutions) with Historically Black Colleges and Universities, Historically Hispanic Colleges and Universities, Tribal Colleges and Universities, and other Minority-Serving Institutions, which have a wealth of knowledge on how to create inclusive environments for learning and career success.

9

- - - - - - - - -

"And So, I'm Going to Work Tomorrow"

Bruce, my friend from our work together in Zaire, has since passed away, so I cannot test any of these ideas on his libertarian sensitivities. If we could share another cold brew now, I would hope that he might be convinced that muddling through the Anthropocene is not a good option and that many of the ideas put forward here, whether as part of a green new deal or under any other name, are not only sensible but essential. While I am not so naïve as to think that Bruce or others who still share his worldviews would be won over easily, I draw inspiration from many others who are leading the way:

- The young African master's student whose poster title I borrowed for chapter 2 inspires me to think about how young scientists all over the world are working for the betterment of their communities and are pressing for structural changes in academia to make science more inclusive.

- The farmers who are adopting regenerative agriculture practice, experimenting and leading by example, and spreading the word to their neighbors of its long-term environmental and economic benefits. A growing number of corporations in food supply chains are backing this trend.
- The early adopters of renewable energy and electric vehicles and the entrepreneurs, scientists, engineers, and companies leading the charge to accelerate these promising trends.
- The schoolchildren who are teaching us as they strike for climate activism, saying that they do not want to be missing school, that they should not be missing school, but that they feel compelled to miss school to send strong and clear signals to grownups that climate change denialism will deny young people the prospect of prosperous, healthy, and equitable futures. These kids and young adults marched outside by the thousands in Glasgow, expressing their deep skepticism that the elders carrying out negotiations inside the COP26 really would deliver what the younger generation knows is needed for their futures. "Prove us wrong!" they pleaded.
- The CEOs and their employees who are committed to reducing their companies' greenhouse gas emissions to targets in alignment with the 2015 Paris Climate Accords while also creating new, high-quality jobs and who have meaningful accomplishments to show for it.
- The asset managers and investors who are taking Environmental-Social-Governance investing seriously, thereby bringing on a sea change in the financial world.
- The millions of people of every race and ethnicity across the world who, following the killing of George Floyd, demonstrated in the streets of their communities for the principles

of justice embodied by the Black Lives Matter movement and the lasting structural changes that these movements are catalyzing.

- The leaders of countries and states who followed the advice of medical science, thereby demonstrating the power of science, engaged with society, to contain the coronavirus pandemic in their communities.

I would also like to share some inspiration from the words of a late friend and colleague, Piers Sellers, who worked as a leading climate scientist for NASA, both before and after becoming an astronaut. Piers died in 2016 from pancreatic cancer at the age of only 61. After receiving his diagnosis but while he still felt well enough to work, he wrote a farewell editorial, in which he confronted what he would do with his remaining days.[1] He did not witness the historic challenges and changes of the early 2020s, but his parting admonishments to us still resonate:

> First, we should brace for change. It is inevitable. It will appear in changes to the climate and to the way we generate and use energy. Second, we should be prepared to absorb these with appropriate sang-froid. Some will be difficult to deal with, like rising seas, but many others could be positive. New technologies have a way of bettering our lives in ways we cannot anticipate. There is no convincing, demonstrated reason to believe that our evolving future will be worse than our present, assuming careful management of the challenges and risks. History is replete with examples of us humans getting out of tight spots. The winners tended to be realistic, pragmatic and flexible; the losers were often in denial of the threat.
>
> As for me, I've no complaints. I'm very grateful for the experiences I've had on this planet. As an astronaut I spacewalked 220 miles above the Earth. Floating alongside the International Space Station, I watched hurricanes cartwheel across oceans, the Amazon snake its way to the sea through a brilliant green carpet of

forest, and gigantic nighttime thunderstorms flash and flare for hundreds of miles along the Equator. From this God's-eye-view, I saw how fragile and infinitely precious the Earth is. I'm hopeful for its future.

And so, I'm going to work tomorrow.

There is reason—good reason—for optimism, but a green new deal, a plan to manage the Anthropocene, or whatever we call it, will not come willy-nilly. It will require the work and inspiration of countless people making their daily choices and daily contributions to a vision of a world in which they and their descendants want to live. And so, we all need to go to work tomorrow, telecommuting or in-person, at our formal jobs or through our efforts as citizens of the planet, and for as many tomorrows as each of us is blessed to enjoy.

ACKNOWLEDGMENTS

I thank John Shordike, Ken Davidson, Brandon Jones, Ray Weil, Jim Williams, Jean Talbert, and two anonymous reviewers for extremely helpful edits and suggestions on drafts of this book. Their contributions greatly improved the final product and challenged me to think deeper and harder on many vexing topics. I thank the editors and staff at Johns Hopkins University Press for taking on this project and supporting it with enthusiasm. I thank scores of colleagues, mentors, and students who have opened my eyes to new ways of viewing topics of sustainability, to doing science, and to communicating to broad audiences through the written word. I am very grateful for the opportunity to serve the American Geophysical Union, where I learned so much about the roles of science in society from very talented, dedicated, and thoughtful staff and fellow volunteer scientists.

Most importantly, I thank my wife, Jean, my son, Bryce, my brother, Ken, and my sister, Wendy, for their support and patience, not only for tolerating me while I sequestered myself in my office to work on "the book" but also for a lifelong journey together of seeking meaning, fulfillment, and joy from our family bonds.

NOTES

Preface

1. US Environmental Protection Agency, "Climate Change and Social Vulnerability in the United States: A Focus on Six Impacts," EPA 430-R-21-003 (September 2021), https://www.epa.gov/system/files/documents/2021-09/climate-vulnerability_september-2021_508.pdf.

2. Vernon R. Morris, "Combating Racism in the Geosciences: Reflections from a Black Professor," *AGU Advances*, 2, e2020AV000358, https://doi.org/10.1029/2020AV000358.

3. Andrea Duane et al., "Towards a Comprehensive Look at Global Drivers of Novel Extreme Wildfire Events," *Climatic Change* 165 (2021): 43, https://doi.org/10.1007/s10584-021-03066-4.

4. Xiaodan Zhou et al., "Excess of COVID-19 Cases and Deaths Due to Fine Particulate Matter Exposure during the 2020 Wildfires in the United States," *Science Advances* 7 (2021): eabi8789, https://doi.org/10.1126/sciadv.abi8789.

5. H. Res. 109, 116th Congress, 1st Session, "Recognizing the duty of the Federal Government to create a Green New Deal," introduced February 7, 2019, https://www.congress.gov/bill/116th-congress/house-resolution/109/text.

6. National Research Council, *Convergence: Facilitating Transdisciplinary Integration of Life Sciences, Physical Sciences, Engineering, and Beyond* (Washington, DC: National Academies Press, 2014), https://doi.org/10.17226/18722.

7. World Meteorological Organization, "State of the Global Climate 2020," WMO-No. 1264, https://library.wmo.int/doc_num.php?explnum_id=10618.

Chapter 1. Muddling or Dealing?

1. James C. Riley, *Rising Life Expectancy* (New York: Cambridge University Press, 2001).

2. FAO, IFAD, UNICEF, WFP, and WHO, *The State of Food Security and Nutrition in the World 2020: Transforming Food Systems for Affordable Healthy Diets* (Rome: FAO, 2020), https://doi.org/10.4060/ca9692en.

3. World Bank, *Poverty and Shared Prosperity 2020: Reversals of Fortune* (Washington, DC: World Bank, 2020), https://doi.org/10.1596/978-1-4648-1602-4.

4. United Nations, Department of Economic and Social Affairs, Population Division, *World Population Prospects 2019: Highlights*, ST/ESA/SER.A/423

(2019) https://population.un.org/wpp/Publications/Files/WPP2019
_Highlights.pdf.

5. Paul J. Crutzen, "Geology of Mankind," *Nature* 415 (2002): 23.

6. Susan Solomon, "The Discovery of the Antarctic Ozone Hole," *Nature* 575 (2019): 46–47.

7. S. Faurby and J.-C. Svenning, "Historic and Prehistoric Human-Driven Extinctions Have Reshaped Global Mammal Diversity Patterns," *Diversity and Distributions* 21 (2015): 1155–1166, https://doi.org/10.1111/ddi.12369.

8. William F. Ruddiman, "The Anthropocene," *Annual Review of Earth and Planetary Sciences* 41 (2013): 45–68, https://doi.org/10.1146/annurev-earth -050212-123944.

9. Colin N. Waters et al., "The Anthropocene Is Functionally and Stratigraphically Distinct from the Holocene," *Science* 351, no. 6269 (2016), https://doi.org/10.1126/science.aad2622.

10. Meera Subramanian, "Humans versus Earth," *Nature* 572 (2019): 168–170, https://media.nature.com/original/magazine-assets/d41586-019 -02381-2/d41586-019-02381-2.pdf.

11. Will Steffen et al., "Planetary Boundaries: Guiding Human Development on a Changing Planet," *Science* 347, no. 6223 (2015), https://doi.org /10.1126/science.1259855.

12. Emily Elhacham et al., "Global Human-Made Mass Exceeds All Living Biomass," *Nature* 588 (2020): 442–444, https://www.nature.com/articles /s41586-020-3010-5.

13. Ellen MacArthur Foundation, *The New Plastics Economy—Rethinking the Future of Plastics* (2017), https://www.ellenmacarthurfoundation.org /publications/the-new-plastics-economy-rethinking-the-future-of-plastics.

14. Jared Diamond, *Collapse* (New York: Penguin, 2005).

15. David Montgomery, *Dirt: The Erosion of Civilizations* (Berkeley: University of California Press, 2007).

16. Will Steffan et al., "Trajectories of the Earth System in the Anthropocene," *Proceedings of the National Academy of Science* 115, no. 33 (2018), www .pnas.org/cgi/doi/10.1073/pnas.1810141115.

17. Intergovernmental Panel on Climate Change, *Climate Change 2021: The Physical Science Basis*, https://www.ipcc.ch/report/sixth-assessment-report -working-group-i/.

18. Adam Smith, *The Wealth of Nations: Books I–III* (London: Penguin, 1999).

19. Heather Boushey, *Unbound: How Inequality Constricts Our Economy and What We Can Do about It* (Cambridge, MA: Harvard University Press, 2019).

20. The Theodore Roosevelt Center, "The Square Deal," https://www .theodorerooseveltcenter.org/Learn-About-TR/TR-Encyclopedia/Politics%20 and%20Government/The%20Square%20Deal.

21. Eric A. Davidson, *You Can't Eat GNP* (New York: Perseus, 2000).

22. Thomas L. Friedman, "The Power of Green," *New York Times Magazine*, April 15, 2007, https://www.nytimes.com/2007/04/15/opinion/15iht-web -0415edgreen-full.5291830.html.

23. Montgomery, *Dirt*.

24. United Nations Department of Economic and Social Affairs, *World Population Prospects 2019: Highlights*, St/ESA/SER.A/423, https://population .un.org/wpp/.

25. National Research Council, *Convergence: Facilitating Transdisciplinary Integration of Life Sciences, Physical Sciences, Engineering, and Beyond* (Washington, DC: National Academies Press, 2014), https://doi.org/10.17226/18722.

26. National Science Foundation, *Growing Convergence Research*, https:// www.nsf.gov/funding/pgm_summ.jsp?pims_id=505637.

Chapter 2. No Tree, No Bee, No Honey, No Money

1. Brittany Goodrich, "Almond Pollination Market: Economic Outlook and Other Considerations," *West Coast Nut*, January 6, 2020, https://www.wcngg .com/2020/01/06/2020-almond-pollination-market-economic-outlook-and -other-considerations/.

2. US Department of Agriculture, National Agricultural Statistics Service, Pacific Region, *California Almond Objective Measurement Report*, July 3, 2019, https://www.nass.usda.gov/Statistics_by_State/California /Publications/Specialty_and_Other_Releases/Almond/Objective -Measurement/201907almom.pdf.

3. Dave Goulson et al., "Bee Declines Driven by Combined Stress from Parasites, Pesticides, and Lack of Flowers," *Science* 347 (2015), doi: 10.1126/ science.1255957; Harry Sivitier et al., "Agrochemicals Interact Synergistically to Increase Bee Mortality," *Nature* 596 (2021): 389–392, doi: 10.1038/s41586- 021-03787-7, https://www.nature.com/articles/s41586-021-03787-7.

4. Eric A. Davidson, *You Can't Eat GNP* (New York: Perseus Publishing, 2000).

5. Thomas Piketty, *Capital in the Twenty-First Century*, trans. Arthur Goldhammer (Cambridge, MA: Belknap Press of Harvard University Press, 2014).

6. Heather Boushey, "The Way We Measure the Economy Obscures What Is Really Going On," *New York Times*, Oct. 28, 2019, https://www.nytimes.com /2019/10/28/opinion/economic-growth-statistics.html.

7. Erin Blakemore, "How the GI Bill's Promise Was Denied to a Million Black WWII Veterans" *History*, June 21, 2019, updated April 20, 2021, https:// www.history.com/news/gi-bill-black-wwii-veterans-benefits.

8. Jeremy S. Homan, Vivek Shandas, and Nicholas Pendleton, "The Effects of Historical Housing Policies on Resident Exposure to Intra-Urban Heat: A Study of 108 US Urban Areas," *Climate* 8, no. 12 (2020) https://doi .org/10.3390/cli8010012; Brad Plumer and Nadja Popovich, "How Decades of Racist Housing Policy Left Neighborhoods Sweltering," *New York Times*, August 24, 2020, https://www.nytimes.com/interactive/2020/08/24/climate /racism-redlining-cities-global-warming.html.

9. X. Wu et al., "Air Pollution and COVID-19 Mortality in the United States: Strengths and Limitations of an Ecological Regression Analysis," *Science Advances* 6, no. 45 (2020): p.eabd4049, https://doi.org/10.1126/sciadv.abd4049.

10. Christopher W. Tessum et al., "PM2.5 Polluters Disproportionately and Systemically Affect People of Color in the United States," *Science Advances* 7 (2021): eabf4491; Mary Angelique G. Demetillo et al., "Space-Based Observational Constraints on NO_2 Air Pollution Inequality from Diesel Traffic in Major US Cities," *Geophysical Research Letters* 48 (2021): e2021GL094333, https://doi.org/10.1029/2021GL094333.

11. Talha Burki, "COVID-19 among American Indians and Alaskan Natives," *The Lancet* 21 (2021): 325–326, https://doi.org/10.1016/S1473-3099(21)00083 -9; Lizzie Wade, "COVID-19 Data on Native Americans Is 'A National Disgrace': This Scientist Is Fighting to Be Counted," *Science* 369 (2020): 1551–1552, https://doi.org/10.1126/science.369.6511.1551.

12. Greg Leiserson, Will McGrew, and Raksha Kopparam, "The Distribution of Wealth in the United States and Implications for a Net Worth Tax," Washington Center for Equitable Growth, March 21, 2019, https:// equitablegrowth.org/the-distribution-of-wealth-in-the-united-states-and -implications-for-a-net-worth-tax/.

13. David Gelles, "C.E.O. Pay Remains Stratospheric, Even at Companies Battered by Pandemic," *New York Times*, April 24, 2021, https://www.nytimes .com/2021/04/24/business/ceos-pandemic-compensation.html.

14. World Food Programme, Food Security Information Network, *Global Report on Food Crisis* (2020) https://docs.wfp.org/api/documents/WFP -0000114546/download/?_ga=2.208511638.54483002.1619610828-2000121269 .1619610828.

15. US Department of Agriculture, Economic Research Service, *Food Security in the U.S.,* last updated September 8, 2021, https://www.ers.usda.gov/topics/food -nutrition-assistance/food-security-in-the-us/key-statistics-graphics.aspx.

16. Heather Boushey, *Unbound: How Inequality Constricts Our Economy and What We Can Do about It* (Cambridge, MA: Harvard University Press, 2019).

17. Shelby R. Buckman et al., "The Economic Gains from Equity," Brookings Papers on Economic Activity, September 8, 2021, https://www.brookings .edu/bpea-articles/the-economic-gains-from-equity/.

18. Samantha Fisher et al., "Air Pollution and Development in Africa: Impacts on Health, the Economy, and Human Capital," *Lancet Planet Health* 5 (2021): e681–e688, https://www.thelancet.com/action/showPdf?pii=S2542 -5196%2821%2900201-1.

19. Katharine Q. Seelye, "Kirk Smith, Towering Figure in Environmental Science, Dies at 73," *New York Times*, June 24, 2020, https://www.nytimes.com /2020/06/24/climate/kirk-smith-dead.html; Kalpana Balakrishnan et al., "In Memoriam: Kirk R. Smith," *Environmental Health Perspectives* 128, no. 7 (July 27, 2020), https://ehp.niehs.nih.gov/doi/10.1289/EHP7808.

20. Tingyu Li et al., "Enhanced-Efficiency Fertilizers Are Not a Panacea for Resolving the Nitrogen Problem," *Global Change Biology* 24 (2018): e511–e521.

21. Al Gore, *An Inconvenient Truth: The Planetary Emergency of Global Warming and What We Can Do about It* (New York: Rodale Books, 2006).

22. Boushey, *Unbound*.

Chapter 3. Are There Too Few or Too Many People?

1. Pierre Friedlingstein et al., "Global Carbon Budget 2019," *Earth System Science Data* 11 (2019): 1783–1838, https://doi.org/10.5194/essd-11-1783-2019.

2. Andrew T. Nottingham et al., "Soil Carbon Loss by Experimental Warming in a Tropical Forest," *Nature* 584 (2020): 234–237, https://doi.org/10.1038/s41586-020-2566-4; Luciana V. Gatti et al., "Amazonia as a Carbon Source Linked to Deforestation and Climate Change," *Nature* 595 (2021): 388–393, https://www.nature.com/articles/s41586-021-03629-6.

3. A. Park Williams et al., "Large Contribution from Anthropogenic Warming to an Emerging North American Megadrought," *Science* 368 (2020): 314–318, https://doi.org/10.1126/science.aaz9600.

4. Jonathan T. Overpeck and Bradley Udall, "Climate Change and the Aridification of North America," *Proceedings of the National Academy of Science* 117 (2020): 11856–11858, www.pnas.org/cgi/doi/10.1073/pnas.2006323117.

5. Jack Healy and Sophie Kasakove, "A Drought So Dire That a Utah Town Pulled the Plug on Growth," *New York Times*, July 20, 2021, https://www.nytimes.com/2021/07/20/us/utah-water-drought-climate-change.html.

6. Rutger Willem Hofste et al., "17 Countries, Home to One-Quarter of the World's Population, Face Extremely High Water Stress," World Resources Institute, https://www.wri.org/insights/17-countries-home-one-quarter-worlds-population-face-extremely-high-water-stress.

7. Tim Searchinger et al., *Creating a Sustainable Food Future* (Washington, DC: World Resources Institute, 2019), https://www.wri.org/research/creating-sustainable-food-future.

8. Emily Sohn, "Planning for Success," *Nature* 588 (2020): S162–S165.

9. Abhijit V. Banerjee and Esther Duflo, *Good Economics for Hard Times* (New York: Public Affairs, 2019).

10. Banerjee and Duflo, *Good Economics for Hard Times*.

11. FSIN and Global Network Against Food Crises, *Global Report on Food Crises, 2021*, https://www.wfp.org/publications/global-report-food-crises-2021.

12. Boyd A Swinburn et al., "The Global Syndemic of Obesity, Undernutrition, and Climate Change: The Lancet Commission Report," *The Lancet Commissions* 393 (2019): 791–846, https://doi.org/10.1016/S0140-6736(18)32822-8.

13. World Disasters Report 2020, *Come Heat or High Water* (Geneva: International Federation of Red Cross and Red Crescent Societies, 2020), https://media.ifrc.org/ifrc/world-disaster-report-2020.

14. Guy J. Abel et al., "Climate, Conflict, and Forced Migration," *Global Environmental Change* 54 (2019): 239–249.

15. IFRC, *Displacement in a Changing Climate* (Geneva: International Federation of Red Cross and Red Crescent Societies, 2021), https://www.ifrc.org/sites/default/files/2021-10/IFRC-Displacement-Climate-Report-2021_1.pdf.

16. Viviane Clement et al., *Groundswell Part 2: Acting on Internal Climate Migration*, (Washington, DC: World Bank, 2021), https://openknowledge.worldbank.org/handle/10986/36248.

17. Elizabeth Svoboda, "Freedom of Choice," *Nature* 588 (2020): S166–S167; Michael Eisenstein, "Stifling Sperm," *Nature* 588 (2020): S170–S171.

18. Sarah G. Chamberlain et al., "Reboot Contraceptives Research—It Has Been Stuck for Decades," *Nature* 587 (2020): 543–545.

19. Bianca Nogrady, "Set and Forget," *Nature* 588 (2020): S168–S169.

Chapter 4. Manure Happens

1. Our World in Data, https://ourworldindata.org/grapher/animals -slaughtered-for-meat?country=~OWID_WRL.

2. United Nations Food and Agriculture Organization, FAOSTAT, http://www.fao.org/faostat/en/#data/EMN. Estimates of global production of livestock manure nitrogen in 2018 from this source were multiplied by a weighted average carbon-to-nitrogen ratio of 15 to estimate total carbon content and then multiplied again by two to estimate total dry mass.

3. Burno Basso et al., "Yield Stability Analysis Reveals Sources of Large-Scale Nitrogen Loss from the US Midwest," *Scientific Reports* 9 (2019): 5774, https://doi.org/10.1038/s41598-019-42271-1.

4. Patricia M. Gilbert, "From Hogs to HABs: Impacts of Industrial Farming in the US on Nitrogen and Phosphorus and Greenhouse Gas Pollution," *Biogeochemistry* 150 (2020): 139–180, https://doi.org/10.1007/s10533-020-00691-6.

5. US Environmental Protection Agency, Mississippi River/Gulf of Mexico Hypoxia Task Force, "Northern Gulf of Mexico Hypoxic Zone," https://www.epa.gov/ms-htf/northern-gulf-mexico-hypoxic-zone.

6. Donald F. Boesch, "Barriers and Bridges in Abating Coastal Eutrophication," *Frontiers in Marine Science* 6 (2019): 123, doi: 10.3389/fmars.2019.00123.

7. Xin Zhang et al., "Managing Nitrogen for Sustainable Development," *Nature* 528 (2015): 51–59.

8. Calvin Harmin, "Flood Vulnerability of Hog Farms in Eastern North Carolina: An Inconvenient Poop" (master's thesis, East Carolina University, 2015), https://thescholarship.ecu.edu/handle/10342/5143.

9. Julia Kravchenko et al., "Mortality and Health Outcomes in North Carolina Communities Located in Close Proximity to Hog Concentrated Animal Feeding Operations," *North Carolina Medical Journal* 79 (2018): 278–288, https://doi.org/10.18043/ncm.79.5.278.

10. Steve Wing et al., "Integrating Epidemiology, Education, and Organizing for Environmental Justice: Community Health Effects of Industrial Hog Operations," *American Journal of Public Health* 98, no. 8 (2008): 1390–1397, https://doi.org/10.2105/AJPH.2007.1104862008.

11. Maria Pfister and Tim Manning, "Stink, Swine, and Nuisance: The North Carolina Hog Industry and Its Waste Management Woes," Environment and Energy Study Institute, August 10, 2018, https://www.eesi.org/articles/view/stink-swine-and-nuisance-the-north-carolina-hog-industry-and-its-waste-mana.

12. Timothy B. Wheeler, "Maryland Firm Plans Chicken Litter Recycling Plant at Perdue's Sussex Facility," *Delmarva Now*, December 9, 2019, https://

www.delmarvanow.com/story/news/local/maryland/2019/12/02/md-firm
-plans-chicken-litter-recycling-plant-perdues-sussex-facility/4295944002/.

13. Aman Azhar, "North Carolina's New Farm Bill Speeds the Way for
Smithfield's Massive Biogas Plan for Hog Farms," *Inside Climate News*,
August 31, 2021, https://insideclimatenews.org/news/31082021/north
-carolina-farm-bill-biogas-smithfield/.

14. US Department of Agriculture, National Agricultural Statistics Service,
2017 Census of Agriculture: Summary and State Data, 1: *Geographic Area Series*,
Part 51, AC-17-A-51, April 2019, https://www.nass.usda.gov/Publications
/AgCensus/2017/Full_Report/Volume_1,_Chapter_1_US/usv1.pdf.

15. Environmental Defense Fund, "Our Partnership with Walmart Brings
Big Change," July 27, 2019, https://www.edf.org/partnerships/walmart.

16. Eric A. Davidson et al., "N-related Greenhouse Gases in North America:
Innovations for a Sustainable Future," *Current Opinion in Environmental Sustainability* 9–10 (2014): 1–8.

17. Peter Newton et al., "What Is Regenerative Agriculture? A Review of Scholar
and Practitioner Definitions Based on Processes and Outcomes," *Frontiers in
Sustainable Food Systems* 4 (2020): 577723. https://doi.org/10.3389/fsufs.2020.577723.

18. Ken E. Giller et al., "Regenerative Agriculture: An Agronomic
Perspective," *Outlook on Agriculture* 50 (2021): 13–25, https://doi.org/10.1177
/0030727021998063.

19. Newton et al., "What Is Regenerative Agriculture?"

20. Gosia Wozniacka, "Carbon Markets Stand to Reward 'No-Till' Farmers:
But Most Are Still Tilling the Soil," *Civil Eats*, May 3, 2021, https://civileats
.com/2021/05/03/carbon-markets-stand-to-reward-no-till-farmers-but
-most-are-still-tilling-the-soil/.

21. Iowa Department of Agriculture and Land Stewardship, "Crop Insurance Discounts Available for Farmers Who Plant Cover Crops," September 30,
2019, https://iowaagriculture.gov/news/crop-insurance-discounts-available
-farmers-who-plant-cover-crops.

22. Gilbert, "From Hogs to HABs."

23. David Leclère et al., "Bending the Curve of Terrestrial Biodiversity
Needs an Integrated Strategy," *Nature* 585 (2020): 551–556, https://doi.org/10
.1038/s41586-020-2705-y.

24. Thomas S. Jayne and Pedro A. Sanchez, "Agricultural Productivity
Must Improve in Sub-Saharan Africa," *Science* 372 (2021): 1045–1047,
https://doi.org/10.1126.abf5413.

25. E. Dinerstein et al., "A Global Deal for Nature: Guiding Principles,
Milestones, and Targets," *Science Advances* 5 (2019): eaaw2869.

26. Ashoka Mukpo, "As COP15 Approaches, '30 by 30' Becomes a Conservation Battleground," *Mongabay*, August 26, 2021, https://news.mongabay
.com/2021/08/as-cop15-approaches-30-by-30-becomes-a-conservation
-battleground/.

27. Callum M. Roberts et al., "Climate Change Mitigation and Nature
Conservation Both Require Higher Protected Area Targets," *Philosophical*

Transactions of the Royal Society B 375 (2020): 20190121, https://doi.org/10.1098/rstb.2019.0121.

28. Richard S. Ostfeld and Felicia Keesing, "Species That Can Make Us Ill Thrive in Human Habitats," *Nature* 584 (2020): 346–347, https://doi.org/10.1038/d41586-020-02189-5.

29. United Nations Environment Programme and International Livestock Research Institute, "Preventing the Next Pandemic: Zoonotic Diseases and How to Break the Chain of Transmission," Nairobi, Kenya, July 6, 2020, https://www.unep.org/resources/report/preventing-future-zoonotic-disease-outbreaks-protecting-environment-animals-and.

30. Andrew P. Dobson et al., "Ecology and Economics for Pandemic Prevention" *Science* 369 (2020): 379–381, https://doi.org/10.1126/science.abc3189.

31. Food and Agriculture Organization of the United Nations, *The State of World Fisheries and Aquaculture 2020* (Rome: FAO, 2020), https://doi.org/10.4060/ca9229en.

32. Rosamond L. Naylor et al., "A 20-Year Retrospective Review of Global Aquaculture," *Nature* 591 (2021): 551–563, https://doi.org/10.1038/s41586-021-03308-6; Jessica A. Gephart et al., "Environmental Performance of Blue Foods," *Nature* 597 (2021): 360–365, https://www.nature.com/articles/s41586-021-03889-2?proof=t.

33. Eric A. Davidson et al., "Nutrients in the Nexus." *Journal of Environmental Studies and Science* 6 (2016): 25–38, 2016, doi: 10.1007/s13412-016-0364-y.

34. Dieter Gerten et al., "Feeding Ten Billion People Is Possible within Four Terrestrial Planetary Boundaries," *Nature Sustainability* 3 (2020): 200–208, https://doi.org/10.1038/s41893-019-0465-1.

35. Cindy Peillex and Martin Pelletier, "The Impact and Toxicity of Glyphosate and Glyphosate-Based Herbicides on Health and Immunity," *Journal of Immunotoxicology* 17 (2020): 163–174, https://doi.org/10.1080/1547691X.2020.1804492.

36. Wozniacka, "Carbon Markets."

37. Eric Sfiligoj, "Herbicide Resistance: The Numbing Numbers from the Weed Wars" *CropLife*, April 2, 2017, https://www.croplife.com/crop-inputs/herbicide-resistance-the-numbing-numbers-from-the-weed-wars/.

38. H. Claire Brown, "Attack of the Superweeds," *New York Times*, August 18, 2021, https://www.nytimes.com/2021/08/18/magazine/superweeds-monsanto.html.

39. Rachel Carson, *Silent Spring* (Boston: Houghton Mifflin, 1962).

40. Surendra K. Dara, "The New Integrated Pest Management Paradigm for the Modern Age," *Journal of Integrated Pest Management* 10, no. 1 (2019): 1–9, https://doi.org/10.1093/jipm/pmz010.

41. Gina Solomon, "The EPA Is Banning Chlorpyrifos, a Pesticide Widely Used on Food Crops, after 14 Years of Pressure from Environmental and Labor Groups," *The Conversation*, August 24, 2021, https://theconversation.com/the-epa-is-banning-chlorpyrifos-a-pesticide-widely-used-on-food-crops-after-14-years-of-pressure-from-environmental-and-labor-groups-166485.

42. Michelle L. Hladik et al., "Environmental Risks and Challenges Associated with Neonicotinoid Insecticides," *Environmental Science and Technology* 52 (2018): 3329–3334, https://doi.org/10.1021/acs.est.7b06388.

43. Rachel Nuwer, "As Locusts Swarmed East Africa, This Tech Helped Squash Them," *New York Times*, April 8, 2021, https://www.nytimes.com/2021 /04/08/science/locust-swarms-africa.html.

44. Brad Spellberg et al., *Antibiotic Resistance in Humans and Animals*, NAM Perspectives Discussion Paper (Washington, DC: National Academy of Medicine, 2016), https://doi.org/10.31478/201606d.

45. National Academies of Sciences, Engineering, and Medicine, *Genetically Engineered Crops: Experiences and Prospects* (Washington, DC: National Academies Press, 2016), https://doi.org/10.17226/23395.

46. Andrew J. Wright, "The Precautionary Tale of Golden Rice," *Science* 366 (2019): 192, https://doi.org/10.1126/science.aaz0466.

47. US Department of Agriculture, Economic Research Service, "Feedgrains Sector at a Glance," updated March 5, 2021, https://www.ers.usda.gov/topics /crops/corn-and-other-feedgrains/feedgrains-sector-at-a-glance/.

48. Henk Westhoek et al., *Nitrogen on the Table: The Influence of Food Choices on Nitrogen Emissions and the European Environment* (Edinburgh: Centre for Ecology and Hydrology, 2015), https://www.pbl.nl/en/publications/nitrogen -on-the-table-the-influence-of-food-choices-on-nitrogen-emissions-and-the -european-environment.

49. Marco Springmann et al., "Options for Keeping the Food System within Environmental Limits," *Nature* 562 (2018): 519–526, https://doi.org/10 .1038/s41586-018-0594-0.

50. M. Crippa et al., "Food Systems Are Responsible for a Third of Global Anthropogenic GHG Emissions, *Nature Food* 2 (2021): 198–209, https://doi.org /10.1038/s43016-021-00225-9.

51. Melissa Clark, "The Meat-Lover's Guide to Eating Less Meat," *New York Times*, December 31, 2019, updated January 2, 2020, https://www.nytimes .com/2019/12/31/dining/flexitarian-eating-less-meat.html.

52. Nitrogen Footprint, http://n-print.org/.

53. Hunt Allcott et al., "Food Deserts and the Causes of Nutritional Inequality," *Quarterly Journal of Economics* 134, no. 4 (2019): 1793–1844, https://doi -org.proxy-um.researchport.umd.edu/10.1093/qje/qjz015.

54. Columbia Climate School / Columbia Water Center, "Punjab, India," https://water.columbia.edu/content/punjab-india.

55. World Bank, *Missing Food: The Case of Postharvest Grain Losses in Sub-Saharan Africa*, report no. 60371-AFR (2021), https://openknowledge .worldbank.org/handle/10986/2824.

56. National Academies of Sciences, Engineering, and Medicine, *A National Strategy to Reduce Food Waste at the Consumer Level* (Washington, DC: National Academies Press, 2020), https://doi.org/10.17226/25876.

57. Mario Loyola, "Stop the Ethanol Madness," *The Atlantic*, November 23, 2019, https://www.theatlantic.com/ideas/archive/2019/11/ethanol-has

-forsaken-us/602191/; Douglas G. Tiffany, "Economic and Environmental Impacts of U.S. Corn Ethanol Production and Use," Federal Reserve Bank of St. Louis, *Regional Economic Development* 5, no. 1 (2009): 42–58, https://core.ac .uk/download/pdf/6755256.pdf.

58. Pedro A. Sánchez, "Tripling Crop Yields in Tropical Africa," *Nature Geoscience* 3 (2010): 299–300, https://doi.org/10.1038/ngeo853.

59. Loyola, "Stop the Ethanol Madness."

60. Do Kyoung Lee et al., "Biomass Production of Herbaceous Energy Crops in the United States: Field Trial Results and Yield Potential Maps from the Multiyear Regional Feedstock Partnership," *Global Change Biology Bioenergy* 10, no. 10 (2018): 698–716, https://doi.org/10.1111/gcbb.12493.

Chapter 5. Climate Change Viewed by a Skeptic at Heart

1. Intergovernmental Panel on Climate Change, *Climate Change 2021: The Physical Science Basis,* https://www.ipcc.ch/report/sixth-assessment-report -working-group-i/.

2. Roland Jackson, "Eunice Foot, John Tyndall, and a Question of Priority," *Notes and Records* 74 (2020): 105–118, https://doi.org/10.1098/rsnr.2018.0066.

3. Eunice Foote, "Circumstances Affecting the Heat of the Sun's Rays," *American Journal of Science and Arts* 22 (1856): 381–382, https://archive.org /details/mobot31753002152491/page/381/mode/2up?view=theater.

4. Leila McNeill, "This Suffrage-Supporting Scientist Defined the Green-house Effect but Didn't Get the Credit, Because Sexism," *Smithsonian Magazine,* December 5, 2016, https://www.smithsonianmag.com/science-nature /lady-scientist-helped-revolutionize-climate-science-didnt-get-credit -180961291/.

5. Piers Forster, "Half a Century of Robust Climate Models," *Nature* 545 (2017): 296–297, https://doi.org/10.1038/545296a.

6. Eric A. Davidson and Marcia K. McNutt, "Red/Blue and Peer Review," *Eos* 98 (August 2, 2017), https://doi.org/10.1029/2017EO078943.

7. Skeptical Science, "Human Fingerprints," https://skepticalscience.com /graphics.php?g=32.

8. Donald J. Wuebbles et al., "Our Globally Changing Climate," in *Climate Science Special Report: Fourth National Climate Assessment,* ed. Donald J. Wuebbles et al., 1:35–72 (Washington, DC: U.S. Global Change Research Program, 2017), https://doi.org/10.7930/J08S4N35.

9. Eric Roston and Blacki Migliozzi, "What's Really Warming the World," *Bloomberg Businessweek,* June 24, 2015, https://www.bloomberg.com/graphics /2015-whats-warming-the-world/.

10. Rebecca Lindsey, "Climate Change: Atmospheric Carbon Dioxide," Climate.gov, https://www.climate.gov/news-features/understanding-climate /climate-change-atmospheric-carbon-dioxide.

11. Jeff Masters, "Reviewing the Horrid Global 2020 Wildfire Season," *Yale Climate Connections,* January 4, 2021, https://yaleclimateconnections.org/2021 /01/reviewing-the-horrid-global-2020-wildfire-season/.

12. National Academy of Sciences, *Climate Change: Evidence and Causes: Update 2020* (Washington, DC: National Academies Press, 2020), https://doi.org/10.17226/25733.

13. American Geophysical Union, *Society Must Address the Growing Climate Crisis Now*, November 2019, https://www.agu.org/Share-and-Advocate/Share/Policymakers/Position-Statements/Position_Climate.

14. American Meteorological Society, *Climate Change*, April 15, 2019, https://www.ametsoc.org/index.cfm/ams/about-ams/ams-statements/statements-of-the-ams-in-force/climate-change1/.

15. World Meteorological Organization, *The State of the Global Climate 2020*, April 21, 2021, https://public.wmo.int/en/our-mandate/climate/wmo-statement-state-of-global-climate.

16. American Society of Agronomy, *Position Statement on Climate Change*, May 11, 2011, https://www.agronomy.org/files/science-policy/asa-cssa-sssa-climate-change-policy-statement.pdf.

17. World Meteorological Organization, *United in Science 2021: A Multi-Organization High-Level Compilation of the Latest Climate Science Information*, https://library.wmo.int/doc_num.php?explnum_id=10794.

18. World Meteorological Organization, *United in Science 2021: A Multi-Organization High-Level Compilation of the Latest Climate Science Information*, https://public.wmo.int/en/resources/united_in_science.

19. Lukoye Atwoli et al., "Call for Emergency Action to Limit Global Temperature Increases, Restore Biodiversity, and Protect Health," *New England Journal of Medicine*, September 5, 2021, https://doi.org/10.1056/NEJMe2113200.

20. Yafang Cheng et al., "Face Masks Effectively Limit the Probability of SARS-CoV-2 Transmission," *Science* 372 (2021): 1439–1443, doi: 10.1126/science.abg6296.

21. *World Disasters Report 2020, Come Heat or High Water* (Geneva: International Federation of Red Cross and Red Crescent Societies, 2020), https://media.ifrc.org/ifrc/world-disaster-report-2020.

22. Sonia Smith, "Unfriendly Climate," *Texas Monthly*, May 2016, https://www.texasmonthly.com/articles/katharine-hayhoe-lubbock-climate-change-evangelist/.

23. Katharine Hayhoe, *Saving Us: A Climate Scientist's Case for Hope and Healing in a Divided World* (New York: Atria, 2022).

24. Adam B. Smith, "2020 U.S. Billion-Dollar Weather and Climate Disasters in Historical Context," *NOAA Climate.gov*, January 8, 2021, updated September 27, 2021, https://www.climate.gov/news-features/blogs/beyond-data/2020-us-billion-dollar-weather-and-climate-disasters-historical.

25. Naveena Sadasivam, "Holding the Bill," *Grist*, March 4, 2020, https://grist.org/climate/insurance-companies-and-lenders-are-responding-to-climate-change-by-shifting-risk-to-taxpayers/.

26. Chuang Zhao et al., "Temperature Increase Reduces Global Yields of Major Crops in Four Independent Estimates," *Proceedings of the National*

Academy of Sciences 114 (2017): 9326–9331, https://doi.org/10.1073/pnas
.1701762114.

27. International Labour Organization, *Working on a Warmer Planet: The
Impact of Heat Stress on Labour Productivity and Decent Work* (Geneva: International Labour Organization, 2019) https://www.ilo.org/wcmsp5/groups
/public/---dgreports/---dcomm/---publ/documents/publication/wcms
_711919.pdf.

28. Douglass Starr, "Fighting Words," *Science* 367 (2020): 16–19.

29. Sadasivam, "Holding the Bill."

30. Kevin Crowley and Akshat Rathi, "Exxon Mobil's Investment Plan
Adds Millions of Tons of Carbon Output, Documents Reveal," *World Oil*,
October 5, 2020, https://www.worldoil.com/news/2020/10/5/exxon-mobil-s
-investment-plan-adds-millions-of-tons-of-carbon-output-documents
-reveal.

31. Swiss Re Institute, *The Economics of Climate Change: No Action Not an
Option*, April 2021, https://www.swissre.com/dam/jcr:e73ee7c3-7f83-4c17
-a2b8-8ef23a8d3312/swiss-re-institute-expertise-publication-economics-of
-climate-change.pdf.

32. Ulrik Boesen and Tom Van Antwerp, *How Stable Is Cigarette Tax
Revenue?*, Tax Foundation, May 4, 2021, https://taxfoundation.org/cigarette
-tax-revenue-tool/.

33. Jidong Huang and Frank J. Chaloupka IV, *The Impact of the 2009 Federal
Tobacco Excise Tax Increase on Youth Tobacco Use*, National Bureau of Economic
Research, April 2012, https://doi.org/10.3386/w18026, https://www.nber.org
/papers/w18026.

34. Citizen's Climate Lobby, *The Basics of Carbon Fee and Dividend*, https://
citizensclimatelobby.org/basics-carbon-fee-dividend/.

35. Gilbert E. Metcalf, *On the Economics of a Carbon Tax for the United States*,
Brookings Papers on Economic Activity, 2019, https://www.brookings.edu
/wp-content/uploads/2019/03/On-the-Economics-of-a-Carbon-Tax-for-the
-United-States.pdf.

36. Sandeep Pai et al., "Meeting Well-Below 2C Target Would Increase
Energy Sector Jobs Globally," *One Earth* 4 (2021): 1026–1036, https://doi.org
/10.1016/j.oneear.2021.06.005.

37. Richard Conniff, "The Political History of Cap and Trade," *Smithsonian
Magazine*, August, 2009, https://www.smithsonianmag.com/science-nature
/the-political-history-of-cap-and-trade-34711212/.

38. Richard Schmalensee and Robert N. Stavins, "Lessons Learned from
Three Decades of Experience with Cap and Trade," *Review of Environmental
Economics and Policy* 11, no. 1 (2017): 59–79, https://doi.org/10.1093/reep/rew017.

39. Cliff Majersik, "What You Need to Know about the Bold New Building
Laws in New York and D.C.," *GreenBiz*, May 14, 2019, https://www.greenbiz
.com/article/what-you-need-know-about-bold-new-building-laws-new-york
-and-dc.

40. Sarah Kaplan and Aaron Steckelberg, "Empire State of Green," *Washington Post*, May 27, 2020, https://www.washingtonpost.com/graphics /2020/climate-solutions/empire-state-building-emissions/.

41. Ramón R. Alvarez et al., "Assessment of Methane Emissions from the U.S. Oil and Gas Supply Chain," *Science* 361 (2018): 186–188, https://doi .org/10.1126/science.aar7204.

42. US Department of State, "United States, European Union, and Partners Formally Launch Global Methane Pledge to Keep 1.5C Within Reach," November 2, 2021, https://www.state.gov/united-states-european-union-and-partners -formally-launch-global-methane-pledge-to-keep-1-5c-within-reach/.

43. David Feldman and Robert Margolis, "Q2/Q3 2020 Solar Industry Update," National Renewable Energy Laboratory, December 8, 2020, NREL/ PR-6A20-78625, https://www.nrel.gov/docs/fy21osti/78625.pdf.

44. Glasgow Financial Alliance for Net Zero, "Amount of Finance Committed to Achieving 1.5°C Now at Scale Needed to Deliver the Transition," November 3, 2021, https://www.gfanzero.com/press/amount-of-finance -committed-to-achieving-1-5c-now-at-scale-needed-to-deliver-the-transition /; Felix Salmon, "The 3 Most Important Wins from COP26 So Far," *Axios*, November 4, 2021, https://www.axios.com/progress-cop26-glasgow-methane -deforestation-8b5ace1f-41e7-4a98-a257-8e560de0e1e4.html.

45. International Energy Agency, *Net Zero by 2050: A Roadmap for the Global Energy Sector* (Paris: IEA, 2021), https://www.iea.org/reports/net-zero-by-2050.

46. National Academies of Sciences, Engineering, and Medicine, *Accelerating Decarbonization of the U.S. Energy System* (Washington, DC: National Academies Press, 2021), https://doi.org/10.17226/25932.

47. James H. Williams et al., "Carbon-Neutral Pathways for the United States," *AGU Advances* 2 (2020): e2020AV000284. https://doi.org/10.1029 /2020AV000284.

48. United Nations Development Programme, *Fossil Fuel Subsidy Reforms* (New York: UNDP, 2021), https://www.undp.org/publications/fossil-fuel -subsidy-reform-lessons-and-opportunities#modal-publication-download."

49. Vincent Gonzales and Lauren Dunlap, "Advanced Nuclear Reactors 101," Resources for the Future, March 26, 2021, https://www.rff.org /publications/explainers/advanced-nuclear-reactors-101/.

50. Mark Betancourt, "Greening the Friendly Skies, *Eos*101 (November 4, 2020), https://doi.org/10.1029/2020EO150975.

51. Pierre Friedlingstein et al., "Global Carbon Budget 2019," *Earth System Science Data* 11 (2019): 1783–1838, https://essd.copernicus.org/articles/11/1783 /2019/.

52. Jane Margolies, "Concrete, a Centuries-Old Material, Gets a New Recipe," *New York Times*, August 11, 2020, https://www.nytimes.com/2020 /08/11/business/concrete-cement-manufacturing-green-emissions.html.

53. Ben Soltoff, "Inside ExxonMobil's Hookup with Carbon Removal Venture Global Thermostat," *GreenBiz*, August 29, 2019, https://www.greenbiz

.com/article/inside-exxonmobils-hookup-carbon-removal-venture-global
-thermostat.

54. Climate Action Tracker, "Glasgow's 2030 Credibility Gap: Net Zero's Lip
Service to Climate Action", November 9, 2021, https://climateactiontracker.org
/publications/glasgows-2030-credibility-gap-net-zeros-lip-service-to-climate
-action/.

55. Bronson W. Griscom et al., "Natural Climate Solutions," *Proceedings of
the National Academy of Sciences* 114 (2017): 11645–11650, https://doi.org/10
.1073/pnas.1710465114.

56. Susan C. Cook-Patton et al., "Mapping Carbon Accumulation Potential
from Global Natural Forest Regrowth," *Nature* 585 (2020): 545–550.

57. Friedlingstein et al., "Global Carbon Budget 2019."

58. Frances Seymour and Nancy L. Harris, "Reducing Tropical Deforestation," *Science* 365 (2019): 756–757.

59. Daniel Nepstad et al., "Slowing Amazon Deforestation through Public
Policy and Interventions in Beef and Soy Supply Chains," *Science* 344 (2014):
1118–1123, https://doi.org/10.1126/science.1248525.

60. Mercedes Bustamante, "Tropical Forests and Climate Change Mitigation: The Decisive Role of Environmental Governance," *Georgetown Journal of
International Affairs*, March 20, 2020, https://gjia.georgetown.edu/2020/03
/20/tropical-forests-climate-change-mitigation-role-of-environmental
-governance/; Xiao Feng et al., "How Deregulation, Drought, and Increasing
Fire Impact Amazonian Biodiversity," *Nature* 597 (2021): 516–521, https://doi
.org/10.1038/s41586-021-03876-7.

61. UN Climate Change Conference UK, "Glasgow Leaders' Declaration
on Forests and Land Use," November 2, 2021, https://ukcop26.org/glasgow
-leaders-declaration-on-forests-and-land-use/.

62. Robin Chazdon and Pedro Brancalion, "Restoring Forests as a Means
to Many Ends," *Science*, 365 (2019): 24.

63. Todd E. Katzner, David M. Nelson et al., "Wind Energy: An Ecological Challenge," *Science* 366 (2019): 1206–1207, https://doi.org/10.1126
/science.aaz9989.

64. Taber D. Allison et al., "Impacts to Wildlife of Wind Energy Siting and
Operation in the United States," *Issues in Ecology*, Report No. 21, Ecological
Society of America, 2019, https://www.esa.org/wp-content/uploads/2019/09
/Issues-in-Ecology_Fall-2019.pdf/.

65. Jim Motavalli, "Soon, the Kitty Litter Will Come by Electric Truck,"
New York Times, August 27, 2020, https://www.nytimes.com/2020/08/27
/business/electric-delivery-vehicles-ups-fedex-amazon.html.

66. Jack Ewing, "World's Largest Long-Haul Truckmaker Sees
Hydrogen-Fueled Future," *New York Times*, May 23, 2021, updated July 21,
2021, https://www.nytimes.com/2021/05/23/business/hydrogen-trucks
-semis.html.

67. George Crabtree, "The Coming Electric Vehicle Transformation,"
Science 366 (2019): 422–424.

68. Bloomberg NEF, "Electric Vehicle Outlook 2020," https://bnef.turtl.co /story/evo-2020/page/1?teaser=yes.

69. Mark A. Andor et al., "Running a Car Costs Much More than People Think—Stalling the Uptake of Green Travel," *Nature* 580 (2020): 453–455.

70. Aakash Arora et al., "Why Electric Cars Can't Come Fast Enough," Boston Consulting Group, April 20, 2021, https://www.bcg.com/publications /2021/why-evs-need-to-accelerate-their-market-penetration.

71. Crabtree, "The Coming Electric Vehicle Transformation."

72. Williams et al., "Carbon-Neutral Pathways for the United States."

73. Climate Action Tracker, "Glasgow's 2030 Credibility Gap."

Chapter 6. The Luddites Had It Half-Right

1. James C. Riley, *Rising Life Expectancy: A Global History* (New York: Cambridge University Press, 2001).

2. Abhijit V. Banerjee and Esther Duflo, *Good Economics for Hard Times* (New York: Public Affairs, 2019).

3. Banerjee and Duflo, *Good Economics for Hard Times*.

4. Sandeep Pai et al., "Meeting Well-Below 2C Target Would Increase Energy Sector Jobs Globally," *One Earth* 4 (2021): 1026–1036, https://doi.org /10.1016/j.oneear.2021.06.005.

5. Eric A. Davidson, *You Can't Eat GNP* (New York: Perseus Publishing, 2000).

6. Klaus Schwab, "The Fourth Industrial Revolution: What It Means and How to Respond, *Foreign Affairs*, December 12, 2015, https://www.foreignaffairs .com/articles/2015-12-12/fourth-industrial-revolution.

Chapter 7. Circular Economies

1. Ellen MacArthur Foundation, *The New Plastics Economy: Rethinking the Future of Plastics* (2016), https://www.ellenmacarthurfoundation.org /publications/the-new-plastics-economy-rethinking-the-future-of-plastics.

2. Robert C. Hale, "A Global Perspective on Microplastics," *Journal of Geophysical Research Oceans* 125 (2020): e2018JC014719, https://doi.org/10 .1029/2018JC014719.

3. Laura Parker, "Plastic Bag Bans Are Spreading: But Are They Truly Effective?" *National Geographic*, April 17, 2019, https://www.national geographic.com/environment/2019/04/plastic-bag-bans-kenya-to-us -reduce-pollution/.

4. Davina Ngei and Amirah Karmali, "Reflecting on Kenya's Single-Use Plastic Bag Ban Three Years On," World Economic Forum, November 25, 2020, https://www.weforum.org/agenda/2020/11/q-a-reflecting-on-kenyas -single-use-plastic-bag-ban-three-years-on/.

5. Hiroko Tabuchi, Michael Corkery, and Carlos Mureithi, "Big Oil Is in Trouble—Its Plan: Flood Africa with Plastic," *New York Times*, August 30, 2020, https://www.nytimes.com/2020/08/30/climate/oil-kenya-africa -plastics-trade.html.

6. Ellen MacArthur Foundation, *The New Plastics Economy*.

7. Pamela L. Geller and Christopher Parmeter, "This Peeler Did Not Need to Be Wrapped in So Much Plastic," *New York Times*, April 5, 2021, https://www.nytimes.com/2021/04/05/opinion/amazon-plastic-waste.html.

8. William A. McDonough, "A Dialogue on Design," *University of Richmond Law Review* 30, no. 4 (1996): 1071–1091, http://scholarship.richmond.edu/lawreview/vol30/iss4/7.

9. William McDonough and Michael Braungart, *Cradle to Cradle: Remaking the Way We Make Things* (New York: North Point Press, 2002), https://mcdonoughpartners.com/cradle-to-cradle-design/.

10. Save our Seas 2.0 Act, Public Law 116-224, December 18, 2020, https://www.congress.gov/116/plaws/publ224/PLAW-116publ224.pdf.

11. Munsol Ju et al., "Solid Recovery Rate of Food Waste Recycling in South Korea," *Journal of Material Cycles and Waste Management* 18 (2016): 419–426, https://doi.org/10.1007/s10163-015-0464-x.

12. George Crabtree, "The Coming Electric Vehicle Transformation," *Science* 366 (2019): 422–424.

13. Benjamin K. Sovacool et al., "Sustainable Minerals and Metals for a Low-Carbon Future," *Science* 367 (2020): 30–33.

14. Ivan Penn and Eric Lipton, "The Lithium Gold Rush: Inside the Race to Power Electric Vehicles," *New York Times*, May 6, 2021, https://www.nytimes.com/2021/05/06/business/lithium-mining-race.html.

15. Cayte Bosler, "Plans to Dig the Biggest Lithium Mine in the US Face Mounting Opposition," *Inside Climate News,* November 7, 2021, https://insideclimatenews.org/news/07112021/lithium-mining-thacker-pass-nevada-electric-vehicles-climate/.

16. Andrew Ross Sorkin, "BlackRock C.E.O. Larry Fink: Climate Crisis Will Reshape Finance," *New York Times*, January 14, 2020, updated February 24, 2020, https://www.nytimes.com/2020/01/14/business/dealbook/larry-fink-blackrock-climate-change.html

17. Gerald Porter Jr., "P&G Holders Rebuke Board with Vote for Deforestation Report," Bloomberg Green, October 13, 2020, https://www.bloomberg.com/news/articles/2020-10-13/p-g-shareholders-vote-in-favor-of-a-deforestation-report

18. Clifford Krauss and Peter Eavis, "Climate Activists Defeat Exxon in Push for Clean Energy," *New York Times*, May 26, 2021, https://www.nytimes.com/2021/05/26/business/exxon-mobil-climate-change.html .

19. CDP Supply Chain Report, "Cascading Commitments: Driving Ambitious Action through Supply Chain Engagement" (2019), https://6fefcbb86e61af1b2fc4-c70d8ead6ced550b4d987d7c03fcdd1d.ssl.cf3.rackcdn.com/cms/reports/documents/000/004/072/original/CDP_Supply_Chain_Report_2019.pdf?1550490556.

20. Net Zero Asset Managers Initiative, https://www.netzeroassetmanagers.org/#.

21. Andrew Ross Sorkin, "BlackRock Chief Pushes a Big New Climate Goal for the Corporate World," *New York Times*, January 26, 2021, https://www

.nytimes.com/2021/01/26/business/dealbook/larry-fink-letter-blackrock
-climate.html.

22. David Gelles and David Yaffe-Bellany, "Shareholder Value Is No Longer
Everything, Top C.E.O.s Say," *New York Times*, Aug. 19, 2019, https://www
.nytimes.com/2019/08/19/business/business-roundtable-ceos-corporations
.html.

23. Peter S. Goodman, "Stakeholder Capitalism Gets a Report Card: It's Not
Good," *New York Times*, September 22, 2020, updated December 2, 2020,
https://www.nytimes.com/2020/09/22/business/business-roudtable
-stakeholder-capitalism.html.

24. Anastasia O'Rourke, "Hypocrisy Is the First Step to Real Change,"
ECOLABELINDEX, March 25, 2008, http://www.ecolabelindex.com/news
/2008/03/25/hypocrisy-is-the-first-step-to-real-change/.

25. Naomi Klein, *On Fire: The (Burning) Case for a Green New Deal* (New
York: Simon and Schuster, 2019).

Chapter 8. Whither the Academy?

1. Rich G. Carter et al., "Innovation, Entrepreneurship, Promotion, and
Tenure," *Science* 373 (2021): 1312–1314, https://www.science.org/doi/10.1126
/science.abj2098.

2. Vernon R. Morris, "Combating Racism in the Geosciences: Reflections
from a Black Professor," *AGU Advances* 2, no. 1 (2021): e2020AV000358,
https://doi.org/10.1029/2020AV000358.

3. Lora Harris et al., "Equitable Exchange: A Framework for Diversity and
Inclusion in the Geosciences," *AGU Advances* 2, no. 2 (2021): e2020AV000359,
https://doi.org/10.1029/2020AV000359; Freeman Hrabowski III and Anthony
Lane, "Sustainability, Inclusive, Excellence, and Our Shared Future," *Green
Schools Catalyst Quarterly* 8, no. 2 (2021): 16–23, https://catalyst.greenschools
nationalnetwork.org/gscatalyst/october_2021/MobilePagedReplica.action?pm
=2&folio=16#pg16.

4. Lauren Lumpkin, "Longtime UMBC President, Who Turned School into
Top Producer of Black Scientists and Engineers, to Retire," *Washington Post*,
August 25, 2021, https://www.washingtonpost.com/education/2021/08/25
/freeman-hrabowski-retire-umbc/.

5. Ernst L. Boyer, *Scholarship Reconsidered: Priorities of the Professoriate*
(Stanford, CA: Carnegie Foundation for the Advancement of Teaching,
1990), https://www.umces.edu/sites/default/files/al/pdfs/BoyerScholar
shipReconsidered.pdf.

6. Cheryl L. B. Manning, "Engaging Communities in Geoscience with
STEM Learning Ecosystems," *Eos* 101 (August 10, 2020), https://doi.org/10
.1029/2020EO147934.

7. Richard Stone, "Indigenous Alaskans Demand a Voice in Research on
Warming," *Science* 369 (2020): 1284–1285.

8. National Academies of Science, Engineering, and Medicine, *Sexual
Harassment of Women: Climate, Culture, and Consequences in Academic Sciences,*

Engineering, and Medicine (Washington, DC: National Academies Press, 2018), https://doi.org/10.17226/24994.

9. National Academies, *Sexual Harassment of Women.*

10. American Geophysical Union, *2017 AGU Scientific Integrity and Professional Ethics,* https://www.agu.org/-/media/Files/Learn-About-AGU/AGU _Scientific_Integrity_and_Professional_Ethics_Policy_document.pdf.

11. Harris et al., "Equitable Exchange."

12. Morris, "Combating Racism."

13. Carter et al., "Innovation, Entrepreneurship, Promotion, and Tenure."

Chapter 9. *"And So, I'm Going to Work Tomorrow"*

1. Piers J. Sellers, "Cancer and Climate Change," *New York Times,* January 16, 2016, https://www.nytimes.com/2016/01/17/opinion/sunday/cancer -and-climate-change.html.

INDEX

acid rain, 111
afforestation, 83–84, 121, 133
Africa, 74; dust storms, 14; food
   wastage, 80–81; propane gas use, 30;
   rural-to-urban transition, 143–44;
   synthetic fertilizer use, 86; water
   shortages, 38–39
aging population, 41–42, 44–45
agricultural extension services, 85–86
agricultural policies: animal waste/
   nutrient management, 54–59;
   conservation policies and, 64–67;
   Green New Deal recommendations,
   132; incentives and subsidies, 54–59,
   84–85; regenerative agriculture, 8,
   62
agriculture: corporate, 54–56; as
   greenhouse gases source, 4–5; linear
   vs. circular design, 167–68; reforestation and, 121–22; technological
   innovations, 144–46, 147, 148;
   workforce, 61–62, 143–46. See also
   industrial agriculture; regenerative
   agriculture
Agronomy Society of America, 93
AGU Advances, 116–17, 121, 129–30
AGU. See American Geophysical Union
air pollution, 24, 118
algal blooms, 53, 55
almond crop, 19–20
Amazon, 128, 165
American Association for the Advancement of Sciences (AGU), 89
American Chemistry Council, 163
American Geophysical Union, 91, 93,
   171
American West, water shortage, 37–39
ammonia gas, 54, 55
Amtrak, 155, 156

animal waste, 50–59, 62, 82, 83;
   circular management system,
   167–68; as environmental pollution,
   54–57; government policies about,
   54–59; recycling into cropland,
   63–64, 76–77; in regenerative
   agriculture, 63–64; synthetic
   fertilizers versus, 52–54, 86
An Inconvenient Truth (Gore), 31
Anthropocene, 1, 4–9, 47, 97–98, 140
antibiotic resistance, 74–75
aquaculture, 67
Arab Spring, 46
Arbery, Ahmaud, 14–15
aridification, 38
Arrhenius, Svante, 89, 90, 92
artificial intelligence (AI), 140, 157
Australia, 113
automation. See technological advances
automobiles, energy efficient. See
   electric-powered vehicles
aviation industry, 119, 120, 133, 135

Banerjee, Abhijit, Good Economics for
   Bad Times, 41, 43–44
bats, 123, 125
battery energy-storage technology, 119,
   127–30, 135, 168
bees and beekeeping, 17–21, 30, 70, 74
Benioff, Marc, 174–75
Biden administration, 14–15, 112–13, 115
biodiversity, 47, 65, 66, 67, 84, 123, 133
biofuels, 56–57, 77, 87, 120
birds, 123, 125–27
Black Americans: economic and social
   disparities, 14–15, 23–24, 33–34,
   46–47; environmental hazard/
   pollution exposures, 55, 56–57
BlackRock, 171–72, 174